REAL SCIENCE
EXPERIMENTS

REAL Science EXPERIMENTS

40 Exciting STEAM Activities for Kids

Jess Harris

Photography by Paige Green

ROCKRIDGE PRESS

Interior and Cover Designer: Lisa Schreiber
Art Producer: Sue Smith
Editor: Eliza Kirby
Production Editor: Mia Moran

Photography © 2019 Paige Green. Styling by Alysia Andriola, pp. ii, vi, 6, 10, 11, 14, 16, 18, 20, 23, 26, 30, 31, 32, 34, 36, 37, 42, 44, 46, 50, 54, 55, 57, 58, 60, 62, 66, 67, 70, 74, 76, 78, 80, 82, 84, 86, 89, 92, 96, 98, 100, 104, 107, 110, 114, 115, 122, 124, 126, 132, 134; Shutterstock pp. 13, 21, 22, 25, 49, 52, 101, 103; NASA/Johns Hopkins University Applied Physics Laboratory/Southwest Research Institute/National Optical Astronomy Observatory, p. 21

Author photo courtesy of Adam Harris

ISBN: Print 978-1-64152-492-6 | eBook 978-1-64152-493-3

R0

To my brother Derrek.

This one's for you.

CONTENTS

Introduction VIII

Chapter 1

How to Use This Book 1

Chapter 2

Science 5

    Insect Mind Control 7

    Static UFO 9

    Seeing the World Through
       a Bubble 12

    Chaos and the Double Pendulum 15

    Bestie Marble Maze Test 17

    Building 3D Lenses 21

    Floating Dollar Bills 24

    Shrinking Plastic 27

    Rotational Races 29

    Rubber Band Power 2-in-1 33

    Dissecting Sound 35

    The Original Fidget Spinner 38

Chapter 3

Technology 41

    Make a Microscope 43

    Secret Patterns of Sound Waves 45

    Light versus Sound 48

    Hydroponic System 51

    Make a Knockoff Ultraviolet
       Light 53

    Super Simple Stethoscopes 56

    Cell Phone Signal Blocking 59

    Homopolar Motor Color Whirl 61

Chapter 4

Engineering 65

   Building with Bioglue 67

   Make Your Own Molds 69

   Walk-Along Spinning Fish 72

   Put a Cork in It 75

   Candle-Powered Boat 77

   Spaghetti Strength 81

   Rock-Hopping Robot 83

   Series versus Parallel String
      Light Circuits 87

Chapter 5

Art 91

   Momentum Drop Painting 93

   Building an Advanced Drawing
      Machine 95

   Chlorophyll Printing 99

   Geological Art 102

   Design a Hologram 105

Chapter 6

Math 109

   Using the Golden Ratio to Make
      a Mechanical Tentacle 111

   Creating Time 113

   Weaving a Magic Square 116

   Icosahedral Virus 118

   Complex Counting Challenge 120

   Ginormous Tetrahedral
      Challenge 123

   Shooting Straws 125

Chapter 7

Putting It All Together 129

   Glossary 130

   For Further Investigation 133

   References 135

   Experiment Index 136

   Index 138

INTRODUCTION

Welcome to *Real Science Experiments: 40 Exciting STEAM Activities for Kids.* This book will guide you through exploration and discovery, which is what science is all about!

This book contains 40 thought-provoking experiments designed for older kids, ages 8 to 12, who are ready for *real* science. These experiments involve connections between the subject areas of Science, Technology, Engineering, Art, and Math (STEAM). The advanced STEAM experiments in this book are for kids who want to take the next step in science learning. Just picking up this book shows you are ready for this challenge!

Let's start with a little background on STEAM. The **S**cience, **T**echnology, **E**ngineering, and **M**ath fields were first grouped together into the acronym STEM because they are important career fields for innovation. Although a lot of attention is given to STEM education, innovation at its core is a creative process. Adding the "A" for **A**rt changed STEM to STEAM and shows the importance of creativity and **design thinking** in real-world projects.

I have been passionate about STEAM my entire life. As a kid, I recorded observations from nature in a science journal. In college, I studied how bacterial flagella play an important part in our immune system.

I spent five years teaching fifth grade and another five years teaching high school science. I am a nationally board certified teacher for—you guessed it—science! Some of my favorite science lessons integrate art, such as watercolor painting with pipettes. I talk all about science on my blog, MrsHarrisTeaches.com. I even named my daughter Ada in honor of Augusta Ada Lovelace, a mathematician from the 1800s whose work provided insight into the potential of computing devices. STEAM *is* my life, and I will show you how it is an integral part of yours!

STEAM is important because it benefits you in school and in everyday life. The experiments in this book rely on all parts of STEAM because understanding STEAM changes the way you think. STEAM is all about connections, and seeing these connections leads to solutions to the world's problems—possibly even problems that do not exist yet! For example, say you are playing outside and notice some burrs stuck on your clothes. You look closely and notice the interlocking hooks that attach the burrs to your shirt. You wonder how this can be used to make a new way to attach things together.

This story is similar to how George de Mestral came up with the invention known as Velcro® in 1941, as described in the Library of Congress article "What is Biomimicry?" Once you start seeing the connections between subjects, you look at everything differently. This is *real* science.

For each experiment in this book, *you* are the scientist. You get to try out new ideas, create amazing results, and sometimes even light things on fire!

Science is the process of *how* knowledge is created. The scientific method is often introduced to children as a simple recipe for conducting experiments, but real science is more complicated. Scientists investigate questions by trying many different things in many different ways. The steps of the scientific method are not a list, but a *process*, as shown in this diagram:

The process of real science does not always follow the steps of the scientific method in the order shown in this circle, but moves forward, stops, and moves backward many times. This book provides support so you can learn to navigate this scientific process on your own.

This book is interactive. From asking questions, making and testing hypotheses, to developing conclusions based on experiments, this book develops the skills used daily by scientists. It goes beyond basic science demonstrations by involving you in the process of real science.

The experiments are educational and build on the scientific principles and concepts you have already learned in school. They reinforce real scientific procedures. Each experiment also includes explanations of scientific words. All the important science words are defined in the glossary at the end of this book.

After each experiment is "The Real Science Behind How and Why" section, which offers more information to deepen your understanding of the concepts in the experiment. The "Now Try This!" section helps you go beyond what you have learned and invent your own science experiments.

These experiments are for older kids to complete on their own, though collaboration with adults or other children is encouraged. Each experiment has easy-to-follow steps that support the process of scientific investigation.

This book helps you be a scientist in your own home using materials you likely already have on hand. Any materials you don't have can be borrowed from a friend or purchased inexpensively at a store. You will not need any specialized science equipment, such as a microscope, for any of these experiments. You can do this!

Now, let's start experimenting!

Cheers,
JESS HARRIS

Chapter 1

How to Use This Book

The acronym for Science, Technology, Engineering, Art, and Math (STEAM) is used to organize this book. Each of the following chapters represents one of the STEAM categories:

Chapter 2: Science

Chapter 3: Technology

Chapter 4: Engineering

Chapter 5: Art

Chapter 6: Math

Each chapter begins with helpful information and is followed by experiments that focus on the specific STEAM subject. Every experiment in this book focuses on at least one of the STEAM subjects, but many experiments are **interdisciplinary**, meaning they combine two or more of

these subjects. Since the concept of STEAM is interdisciplinary, so are these experiments. To help show these connections, each experiment has a color-coded tab showing other STEAM subjects that match the experiment. At the end of this book, the Experiment Index (page 136) shows *all* the experiments that match each category. Many experiments appear in multiple categories.

GETTING READY

This book is full of fun and interesting experiments that cover a wide variety of topics. Each experiment is designed to capture your interest and build curiosity. All aspects of STEAM are fueled by curiosity, and by the excitement over not knowing something. Look through the table of contents and flip through the pages to find which experiments most intrigue you. Remember, there is no need to start at the beginning and complete the experiments in order. Follow the science journey that is right for *you*.

Keeping a science notebook is an important part of this journey. Many famous scientists and inventors—including Albert Einstein, Nikola Tesla, and Leonardo da Vinci—kept detailed notebooks of their work. Writing in notebooks is not just limited to science. Famous artists, writers, and presidents have been known to journal. Use your science notebook to write and sketch important thoughts as you try out each experiment. Even if you have not yet completed an experiment, you can still collect research in your notebook about science experiments in this book!

FOLLOW SAFETY RULES

One of the most important steps in doing any experiment is thinking about any possible safety risks and making plans to stay safe. Read and follow the cautions for each experiment. An experiment might require adult help, goggles, or other special safety precautions. Here are some basic safety rules to follow when doing science:

1. Follow directions carefully.

2. Tell an adult which experiment you are doing, even if the experiment does not require adult help.

3. Follow all cautions given in the experiment.

4. Dress appropriately by wearing clothing and shoes that help you stay safe. Pull back long hair and wear closed-toe shoes.

5. Ask an adult for help if you feel unsure or unsafe about any part of an experiment.

When you find an experiment you want to do, read through the instructions completely. Make sure the level of difficulty matches your skill level and confidence. Also, check to make sure that you have enough time to complete the experiment. Many of the experiments can be completed in less than an hour, but a few take a lot more time from start to finish. Carefully look over the materials list and gather everything you will need for the experiment.

Read "The Real Question." This section gives you a peek at the experiment, but most importantly, it asks you to form a **hypothesis** to write in your notebook. A hypothesis is a temporary and testable suggested answer to a question. The experiment should give you information to help prove or disprove your hypothesis. For example, one of the experiments in this book investigates the strength of different glues. Your hypothesis for this experiment might be, "Frosting will be a stronger glue than marshmallow cream because it spreads easily and completely on the graham cracker." The experiment will give you evidence to support or change your hypothesis.

Do not worry if you don't feel confident about your hypothesis. It is only based on the information you know right now, and you are learning new things every day. Because a hypothesis is temporary, you update it as you learn more information while conducting your experiment. Don't erase your first hypothesis. Simply draw a line through it, and write a note about why you believe it is wrong. Then write the new one. Having a wrong hypothesis is just as valuable as having a correct one, because *knowing what doesn't work is as important as knowing what does*.

DOING THE EXPERIMENT

Now it is time to *do* to the experiment! Carefully follow the step-by-step instructions. If you are making changes because you are redoing an experiment, record any of these changes in your notebook as well. Measurements in each experiment are given in customary units followed by metric units, often in parentheses. It is important to be familiar with the metric system because it is the standard system of measurement in science. The approximate symbol (≈) is used when measurements are similar but not the same.

Stay positive, even if the experiment does not work the way you thought it would. *Failure is an important part of being a scientist.* We often learn more from failure than we do when everything works perfectly. Failure can be frustrating and difficult, but this is how you learn and grow.

As soon as the experiment is complete, write down all your observations. Include answers to the questions from the "Observations" section.

Was your hypothesis supported? If not, how would you change your hypothesis? What was interesting or unexpected about the experiment? Why do you think things turned out the way they did?

The scientific terminology used in this book is important to know. Bolded terms are defined in the glossary located near the end of the book. Use this scientific terminology when you write in your science notebook.

For each experiment, be sure to read "The Real Science Behind How and Why." This section explains the science topics related to the experiment. Reading it will deepen your understanding and may inspire you to research related topics to learn even more.

Remember, *you* are the scientist! Check out the "Now Try This!" section for ideas to redesign the experiment. This section provides tips to take the experiment one step or even ten steps further. This is where the magic of learning happens. You get to practice using your skills as a scientist by asking new questions and completing your own experiments.

Redoing an experiment or conducting a similar experiment is like listening to your favorite song. Even though you have listened to the same song one thousand times, sometimes you may notice something new, like a bass line or a harmony in the chorus. When you revisit the same science topic, even though the topic has not changed, your perspective and understanding have grown.

Chapter 2

Science

This chapter focuses on exploring big ideas in science, from the physics of motion to the biology of insects. The experiments will let you control ladybugs, see the world through a bubble, make money seemingly levitate, and more—all through the power of science!

Some of the experiments are really two experiments in one. The Dissecting Sound (page 35) device can double as a musical instrument. The Rubber Band Power 2-in-1 (page 33) experiment creates a mechanical robot that can move on its own! Think creatively about the other experiments to figure out how to turn them into other things.

You probably have all the supplies needed for the experiments in this chapter, but there are three things you may have to buy. Depending on where you live and the time of year, it may be hard to find insects for the Insect Mind Control experiment (page 7). Native ladybugs are best for this and can be purchased inexpensively online or at a local garden store. For the Chaos and the Double Pendulum (page 15) experiment, you will need two fidget spinners, which you can buy from a dollar store. For the Dissecting Sound experiment, you may choose to use PVC pipe instead of cardboard tubes. You will need at least 7½ feet (≈2.2 meters) of 1-inch (≈2.5-cm) diameter PVC pipe and a tool to cut PVC pipe, which can be bought cheaply at a hardware store. Some hardware stores will cut the PVC pipe to size for you!

As you become better at scientific thinking, you need to build your understanding of one very important word—variable. A **variable** is something that changes. For example, let's say you let a container full of water sit out for a couple of days. The water slowly evaporates until the container is empty. Variables in this example might be the temperature of the water, the temperature of the air, the volume (amount) of water in the container, or the rate at which the water evaporates. Variables can be measured using numbers or described by qualities such as color or shape. For each experiment, record all the possible variables in your science notebook.

Use the "Getting Ready" and "Doing the Experiment" sections in chapter 1 as a guide. Now, let's experiment!

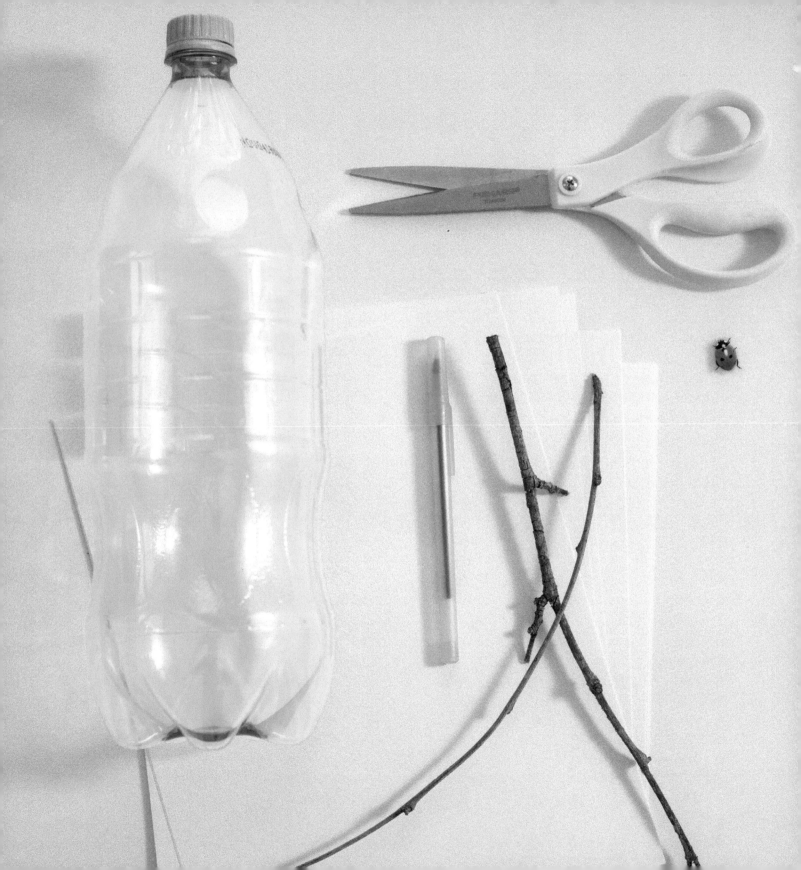

INSECT MIND CONTROL

LEVEL OF DIFFICULTY: EASY

FROM BEGINNING TO END: 45 MINUTES

OTHER CATEGORIES: ENGINEERING

THE REAL QUESTION:

How can you control the motion of an insect using a line drawn on paper with an ink pen? Form a hypothesis about how you think the bug will react. Learn about **pheromones** and insect behavior with this easy experiment. Practice skills used by **entomologists**.

⚠ CAUTION: Have an adult join you on your bug hunt and help you with this experiment. Some insects may bite or cause allergic reactions. Handle live insects carefully.

MATERIALS:

- ⊝ **2-liter soda bottle (clean and empty)**
- ⊝ **Scissors**
- ⊝ **Ladybug, ant, or termite**
- ⊝ **Stick**
- ⊝ **Paper**
- ⊝ **Ballpoint pen with blue ink**

THE STEPS:

1. Create your insect trap. Cut the soda bottle in half right where it starts to curve at the top. Flip the top half of the bottle upside down and put it inside the bottom piece so it looks like a funnel.

2. Find your insect outside with an adult. Allow the insect to crawl onto a stick. Gently tap the stick on the side of your insect trap until the insect falls inside the bottle.

3. Draw some lines onto your paper using the blue ballpoint pen. Fresh, wet ink works best.

4. Take the funnel off the insect trap. Quickly flip the bottom of the bottle over and place it over the paper. Give the insect some time to calm down before observing its behavior.

5. Observe the insect. If needed, gently tap the side of the container so that the insect will move onto the paper.

6. When finished, safely release your insect back into nature.

OBSERVATIONS:

- ⊝ **How did the insect react to the lines you drew on the paper?**

CONTINUED ➜

The Real Science Behind How and Why: Most insects cannot hear or see very well. This means they use other senses to find their way. Some insects do this with **chemoreceptors** in their antennae. Chemoreceptors sense chemicals called pheromones. Insects make and use different pheromones for different reasons. Pheromones can help insects avoid danger or find their way home. When insects find a pheromone trail, they will avoid or follow the path. They "sniff" their way using their antennae. The ink of a blue ballpoint pen has a chemical similar to a pheromone made by insects such as ladybugs, ants, and termites.

"CHEMO-" MEANS CHEMICAL

Now Try This! *What happens if you draw circles or zigzags instead of straight lines? What happens if you use different brands and types of pens or different colors of ink?*

STATIC UFO

LEVEL OF DIFFICULTY: EASY

FROM BEGINNING TO END: 10 MINUTES

OTHER CATEGORIES: MATH

THE REAL QUESTION:

How can **electrostatic force** be used to float an object? Form a hypothesis about how you can use a balloon to float a ring of plastic. Create a floating ring and learn about electrostatic force.

MATERIALS:

- Scissors
- Plastic bags (thin grocery bags work best)
- Tape
- Balloon
- Wool material (blanket, scarf, sweater, or ball of yarn)
- Stopwatch

THE STEPS:

1. Cut out three thin strips of plastic from the bags: one that is 1 inch (≈2.5 cm) wide by 8 inches (≈20 cm) long, one that is 1 inch (≈2.5 cm) wide by 12 inches (≈30 cm) long, and one that is 1 inch (≈2.5 cm) wide by 16 inches (≈41 cm) long.

2. Individually tape each strip to itself in a loop, creating three plastic rings.

3. Blow up a balloon and tie it closed.

4. Rub the balloon and one of the plastic rings with the wool material. If you don't have any wool, you can rub the balloon and rings on your hair instead.

5. Hold the plastic ring with one hand and balloon with the other hand. Slowly move the ring over the balloon and let go of the ring as it begins to hover. Move the balloon as needed to keep the plastic ring above it. Use the stopwatch to see how long you can keep the ring in the air.

6. Repeat this process with the other two plastic rings.

CONTINUED ➔

OBSERVATIONS:

- How long can you get each plastic ring to hover? Which plastic ring is easiest to hover?

The Real Science Behind How and Why: Rubbing wool on both the plastic and the balloon removes electrons from the wool, giving the plastic and the balloon an excess of electrons, or a negative charge. Charges of the same type repel each other. Because the plastic ring is very light, the electrostatic force is strong enough to overcome the force of gravity and physically move it.

Now Try This! *Create additional rings of different widths and lengths. How do the different sizes' float times compare? Can you get other materials to float, such as thin sheets of plastic packing foam? Which material floats the easiest and longest?*

SEEING THE WORLD THROUGH A BUBBLE

LEVEL OF DIFFICULTY: EASY

FROM BEGINNING TO END: 15 MINUTES

OTHER CATEGORIES: ENGINEERING

THE REAL QUESTION:

How do light waves make different colors appear on the surface of a bubble? Form a hypothesis about what colors you will see in a bubble film and why. Create your own bubble solution and loop wand to see the world through a bubble!

! CAUTION: Avoid getting bubble solution in your eyes or mouth.

MATERIALS:

- Corn syrup or sugar
- Warm water (distilled water is best, but tap water will work)
- Baking powder
- Dish soap
- Large shallow container (baking pan or baking sheet with an edge)
- String (cotton twine or yarn)
- 2 straws

THE STEPS:

1. Create your bubble solution. Add 2 teaspoons (10 ml) of corn syrup or ½ cup of sugar (100 grams) to 4 cups (≈1 liter) of warm water. Gently stir in 1 tablespoon of baking powder (≈14 grams) and ½ cup (≈120 ml) of dish soap. You can make your bubble solution directly in your large shallow container and then store it in a cup. Storing your bubble solution for at least an hour before using it will create better bubbles.

2. Create your bubble wand. Thread 3 feet (≈1 m) of string through the two straws and tie to create a loop. Pull the string so the knot is hidden inside one of the straws.

3. Wet one hand with the bubble solution and hold your bubble loop with the wet hand on one straw. Dip your entire bubble wand loop into the bubble solution. Slowly pull the loop up and out of the solution. Look through your bubble film!

OBSERVATIONS:

- What colors do you see in the bubble sheet? How long does the bubble sheet last until it pops? Try to put your other hand through the bubble sheet, first using a dry hand and then after dipping your hand in bubble solution. What other items can you put through the bubble sheet?

Now Try This! *How does adding food coloring to your bubble mixture change the colors you see? Create a bubble loop wand using a longer string. What is the largest bubble sheet you can make? Create your own bubble recipe using other ingredients such as cornstarch, glycerin, and different brands of dish soap. Design an experiment to find out which recipe creates the longest-lasting bubbles.*

The Real Science Behind How and Why:
The primary colors of light are different than the primary colors of paint. When you mix paint, you get black from a subtractive mixture of the primary paint colors blue, red, and yellow. When you mix colors of light, you get white from an additive mixture of the primary colors of light, which are red, green, and blue. The alternating bands of color on the soap film of the bubble sheet are made by the reflection and interference of light waves. Just like ocean waves, light waves have peaks (crests) and valleys (troughs). Destructive interference happens when the crest of one light wave meets the trough of another. They cancel each other out. As the thickness of the soap film changes, it can cause destructive interference of one of the primary colors, canceling out that color. As a result, you only see the two

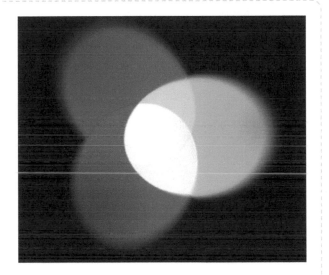

remaining colors. For example, when there is destructive interference of green light waves, you will see a mixture of the two remaining primary colors (red + blue = purple). When you do not see any color in the bubble film, that means it is so thin it's about to pop!

CHAOS AND THE DOUBLE PENDULUM

LEVEL OF DIFFICULTY: EASY

FROM BEGINNING TO END: 15 MINUTES

OTHER CATEGORIES: MATH

THE REAL QUESTION:

Can you predict the motion of a double pendulum? Form a hypothesis about how a double pendulum will move when released from different positions. Build a double pendulum and learn about **chaos theory**.

! CAUTION: Ask an adult for help with the hot glue gun.

MATERIALS:

- Hot glue gun and glue sticks
- 2 fidget spinners
- 3 wooden jumbo size craft sticks, about 6 inches (≈15 cm) x ¾ inch (≈19 mm) in size
- Packing tape

THE STEPS:

1. Add a pea-sized amount of glue to the center disc of one fidget spinner and quickly press one end of a craft stick flat into the glue. Hold the craft stick in place with pressure until the glue cools.

2. On the opposite side of the fidget spinner, add a pea-sized amount of glue to any one of the arms and quickly press one end of the second craft stick flat into the glue. As you construct the double pendulum, each step will build on the new level so the pendulum will move freely and not get stuck on itself.

3. Add a pea-sized amount of glue to the center disc of the second fidget spinner and quickly press the free end of the second craft stick flat into the glue.

4. On the opposite side of the second fidget spinner, add a pea-sized amount of glue to any one of the arms and quickly press one end of the third craft stick flat into the glue.

5. Tape the first craft stick securely to the edge of table so that the double pendulum is hanging off the side.

CONTINUED →

6. Hold the two bottom craft sticks 90 degrees to the top stick (parallel to the floor). Let go and observe the motion of the pendulum. It will swing or loop (make a complete circle). Repeat this several times and record the motion. Use a coding system to record the motion in your science notebook so you can accurately describe it. Use "LR" for when it loops right, "LL" for looping left, "R" for when it swings right, and "L" for when it swings left.

OBSERVATIONS:

○ **What patterns can you find in the pendulum's motion? How does the motion change when you release the bottom two craft sticks from different positions?**

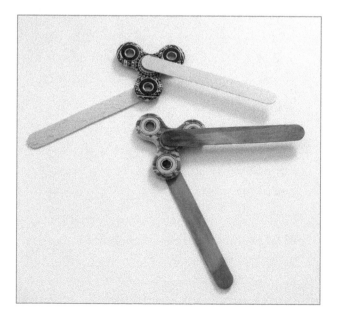

The Real Science Behind How and Why: The pendulum you created in this experiment is a double-rod pendulum, which is one of the simplest examples involving **dynamics** that creates chaotic motion. Small variations in the way the double pendulum starts become amplified as it moves. These changes cause chaotic behavior that shows the **butterfly effect**. The butterfly effect is an example of chaos theory, which is a part of mathematics that looks at systems that are very sensitive, in which a very small change may result in a completely differently outcome.

Now Try This! *Turn a fidget spinner into an* **anemometer** *to measure wind speed. Create the sails out of three small squares of paper about 2 inches (5 cm) x 4 inches (10 cm) in size or use three halves of the smaller side on plastic eggs. Hot glue one sail on each arm of the fidget spinner.*

BESTIE MARBLE MAZE TEST

LEVEL OF DIFFICULTY: HARD

FROM BEGINNING TO END: 90 MINUTES

OTHER CATEGORIES: ENGINEERING, ART

THE REAL QUESTION:

How can you design a collaborative maze? Form a hypothesis about how the time to complete the maze will change over time and why. Test your collaboration with a partner and experiment with gravity at the same time.

MATERIALS:

- Paper and pencil
- Large shallow container (baking sheet with an edge, or a shoebox lid)
- Scissors
- Masking tape
- Straws or clay
- Marble
- 12 unsharpened pencils or wooden dowels
- 8 binder clips
- String

THE STEPS:

1. Use the paper and pencil to sketch out the design of your marble maze. Be sure to have a beginning and end.

2. Construct the maze on your base using your chosen building materials. Either cut and tape down the straws or roll and press down clay to form your marble maze layout. Make sure the path is wide enough for your marble to roll freely. Do test runs using the marble as you build your maze. Make adjustments as you go along.

3. Tape three pencils together to create a tripod tower. Do this again three more times until you have four towers: one for each corner of the marble run base. (These towers will be used to guide the marble through the maze.)

4. Attach one binder clip to the top of each tripod tower with the loop of the clip pointing up. Add tape to add make sure the binder clip does not move or come off.

5. Place each tower about 2 inches (5 cm) diagonally from the four corners of the marble maze base. Tape the tower to the surface it is standing on to add stability.

CONTINUED →

6. Cut four pieces of string, each about 16 inches (≈41 cm) in length. Attach one string to each corner of the marble run base. Run the other end of the string through the loop of the binder clip. Tie the loose end of the string to another binder clip so that it hangs about halfway down.

7. Test out your marble maze with a partner. Each person will hold one hanging binder clip in each hand and slowly pull to tilt the base and direct the marble through the maze.

OBSERVATIONS:

● Record the amount of time it takes to complete the maze. Calculate the average time to complete the maze by adding the three best times and dividing by three. What conditions led to the quickest and slowest times?

The Real Science Behind How and Why: Put your hand on the back of your head, right above your neck. There, inside your skull, is the part of your brain called the cerebellum, which controls your hand-eye coordination. Deeper inside is a part of your brain called the hippocampus, which is important for **spatial navigation.** Scientists study how rats react in mazes to learn more about these parts of the brain. Research the differences between mazes and labyrinths to discover how mathematical algorithms can be used to make and solve these puzzles.

Now Try This! *Have different people try the maze. Which pair of people can complete the maze the quickest? How do times compare if pairs are allowed to talk versus staying silent?*

Does the average completion time change for different times of the day? Can you complete the maze by yourself using your hands and feet?

BUILDING 3D LENSES

LEVEL OF DIFFICULTY: EASY

FROM BEGINNING TO END: 15 MINUTES

OTHER CATEGORIES: TECHNOLOGY, ENGINEERING, ART

THE REAL QUESTION:

How do three-dimensional (3D) images work? Form a hypothesis about how red and blue lenses affect vision. Build 3D lenses to experiment with vision.

 CAUTION: Looking through 3D lenses for too long can cause eye strain.

MATERIALS:

- 3-inch (~7.5-cm) strip of plastic wrap
- Red permanent marker
- Blue permanent marker
- Scissors
- Ruler
- Thin cardboard (thickness of a cereal box)
- Transparent tape
- White paper
- Pencil
- Blue highlighter marker
- Pink highlighter marker

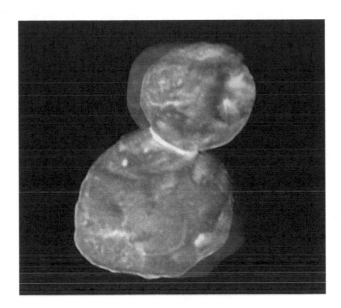

This image of Arrokoth can be viewed with red-blue 3D lenses to reveal the Kuiper Belt object's three-dimensional shape.

THE STEPS:

1. Lay the plastic wrap flat on a table.

2. Color a 2- by 2-inch (5- by 5-cm) square solid red on the plastic wrap using a permanent marker. Color another 2- by 2-inch (5- by 5-cm) square using the blue permanent marker. Allow both to dry. These will be your color filters.

CONTINUED →

3. Cut your cardboard into two rectangles, each 4 inches (≈10 cm) by 2 inches (≈5 cm) in size. These will be your 3D lenses.

4. On one short side of each rectangle, cut out a 1.2- by 1.2-inch (3- by 3-cm) square (so it looks like someone took a bite out of the short edge of the card).

5. Cut out the red and blue color filters you made.

6. Carefully tape the filters as flat as possible over the 1.2- by 1.2-inch (3- by 3-cm) square holes in your cardboard. You want to avoid wrinkles.

7. Now, draw your own simple line drawing on the white paper, first using pencil. Choose something you can draw easily, like a car or a smiley face. Next, add a blue line a few millimeters to the left of the pencil line using the blue highlighter (like a shadow). Then, add a pink line a few millimeters to the right of the pencil line using the pink highlighter.

8. Hold the blue filter over your right eye and the red filter over your left eye and look at your pink and blue 3D picture. This image is called an **anaglyph**.

OBSERVATIONS:

⊙ **What do you see when you look only out of your right or left eye while holding one filter over each eye?**

The Real Science Behind How and Why:
Two factors come into play to make this work. First, the color filters only allow the opposite color to show. You can see this for yourself by drawing a horizontal line with pink and a vertical line with blue, then closing one eye at a time when viewing them through the filters. Your right eye will only see what is drawn in pink and your left eye will only see what is drawn in blue.

The second factor is that your eyes are actually seeing two different things all the time. Each eye has its own perspective. You can see this for yourself by holding your pointer finger at arm's length in front of you. Close your right eye and notice where the image of your finger appears to be located in the background of the room. Without moving your finger, close your left eye. Now the image of your finger appears to have moved to a different spot. The horizontal distance between the two perspectives of your finger is an example of **parallax**. Your brain calculates how far objects are away from you by using this parallax. When you draw a pink and blue picture and then wear filters over your eyes, your brain is tricked into thinking your flat picture is three-dimensional by presenting each eye with a slightly different version of the picture.

Now Try This! *Create additional drawings with different amounts of space between the original pencil line and the pink and blue highlighters. Which spacing makes objects appear closer or farther away?*

FLOATING DOLLAR BILLS

LEVEL OF DIFFICULTY: EASY

FROM BEGINNING TO END: 15 MINUTES

OTHER CATEGORIES: ART, MATH

THE REAL QUESTION:

The center of gravity is the balance point of an object. How does the shape and weight distribution of an item influence its center of gravity? Form a hypothesis about the location of the center of gravity for different objects. Learn how to balance a dollar bill to make it look like it is floating.

! CAUTION: Do not balance breakable items.

MATERIALS:

- Items to balance (marker, ruler, pencil, pencil with different amounts of clay attached to each end, etc.)
- Dollar bill
- Nickel

THE STEPS:

1. Practice finding the center of gravity of different items. This is the point where you can balance the item using just one finger. Hold the item horizontally. Place each pointer finger under the two ends of the item. Slowly and gently, slide your fingers together. The location where your fingers meet is the center of gravity for that item. At this spot, you can balance the item with only one finger.

2. Fold a dollar bill in half the long way.

3. Place a nickel inside the first fold, and fold again the long way. The second fold cannot be perfectly in half because of the size of the nickel, so fold it to the diameter of the nickel.

4. Make sure the nickel is all the way to one of the far ends of the dollar bill without sticking out. Put one of your pointer fingers under each end of the dollar bill. Slowly move your finger on the non-nickel side out from under the bill. Once you find the center of gravity on the nickel side, the dollar bill will balance. Because the center of gravity is so far from the center, the dollar bill will look like it is floating. This will take many attempts, so keep trying!

5. Your fingers will likely need to be about a half inch (1 to 2 cm) from the ends of the dollar bill rather than being directly on the edge. To maintain the illusion, secretly hide the nickel in your hand before handing the dollar bill over for others to try to balance.

OBSERVATIONS:

- How does the location of the center of gravity relate to the distribution of mass of an item?

The Real Science Behind How and Why:
If an object is symmetrical and evenly weighted, its center of gravity will actually be the center of the object. If an object is irregular in shape, its center of gravity will be closer to its heavier end. The terms "center of gravity" and "center of mass" can be used interchangeably. In astronomy, the center of mass between two celestial bodies (like the Earth and sun) is called the barycenter. The barycenter is the point around which the objects orbit. Just like center of mass, the barycenter is closest to the object with the most mass. You can research to learn how barycenters help astronomers discover new planets.

Now Try This! *Try finding the center of gravity of items when they are vertical instead of horizontal. Try finding the center of gravity of irregularly shaped items, like an empty soda bottle. Add different amounts of water to the soda bottle to investigate the impact on finding the center of gravity.*

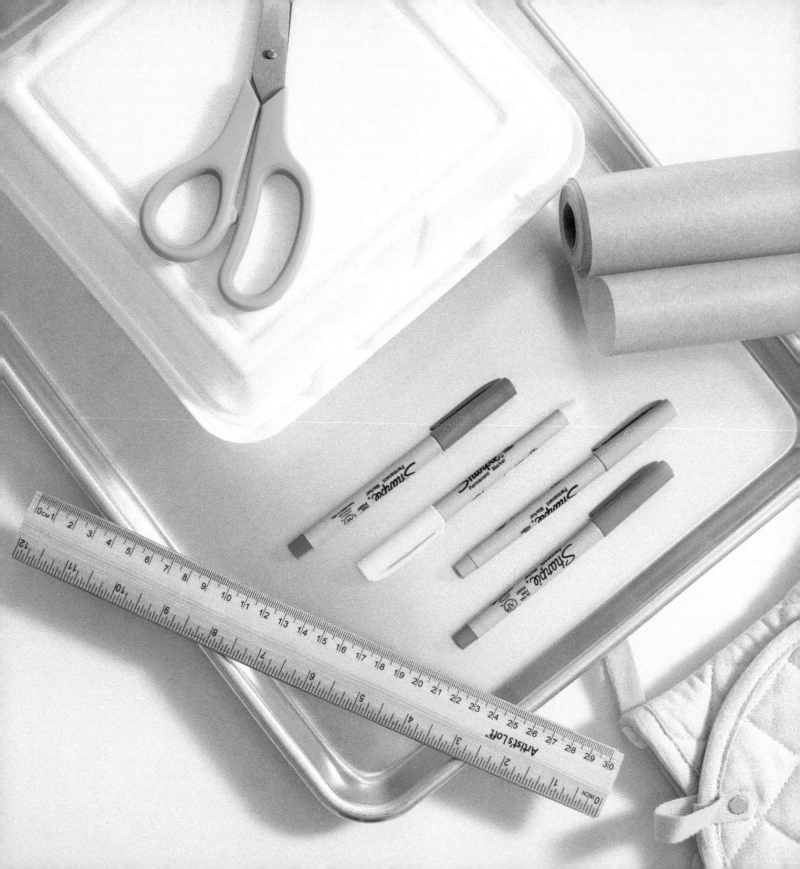

SHRINKING PLASTIC

LEVEL OF DIFFICULTY: EASY

FROM BEGINNING TO END: 30 MINUTES

OTHER CATEGORIES: TECHNOLOGY, ART, MATH

THE REAL QUESTION:

How can plastic shrink? Form a hypothesis about how much heated plastic shrinks after it cools. Practice skills used by chemical engineers and learn about **thermoplastic**.

! CAUTION: Have an adult help you use the oven. Carefully handle hot items.

MATERIALS:

- ⊘ **Permanent markers**
- ⊘ **Polystyrene (plastic stamped with a #6 such as restaurant takeout container or egg carton)**
- ⊘ **Oven**
- ⊘ **Ruler**
- ⊘ **Metal baking sheet**
- ⊘ **Aluminum foil or parchment paper**
- ⊘ **Spatula**
- ⊘ **Oven mitt**

THE STEPS:

1. Using permanent markers, color some designs on the plastic and then cut out each design. Many designs work well, from unicorns to emojis, so use your imagination!

2. Preheat the oven to 325 degrees Fahrenheit (≈163 degrees Celsius).

3. Measure and record the length and width of each plastic piece.

4. Place the plastic designs on a metal baking sheet covered with aluminum foil or parchment paper.

5. Put in the oven and bake the plastic for 3 to 7 minutes. While baking, the plastic will first curl up and then flatten. Once this happens, have an adult remove the baking sheet from the oven with an oven mitt. When the plastic is first removed from the oven, you can use a spatula to flatten the piece if necessary.

6. Measure and record the new length and width of each piece.

OBSERVATIONS:

- ⊘ **How does the shape and size of each piece compare before and after baking?**

CONTINUED ➔

The Real Science Behind How and Why:
Polystyrene is one of the most commonly used plastics. Polystyrene is made of long chains of molecules, called polymers, which are all jumbled together. During the manufacturing process, polystyrene is heated and stretched out into thin sheets. The plastic is cooled quickly to keep its thin shape, but the polymers prefer their original jumbled arrangement. When you heat polystyrene in the oven, the polymer chains shrink back to their original shape. Polystyrene is not biodegradable and, as a result, some countries have banned using it in takeout containers.

Now Try This! *Self-folding plastics is a recent area of scientific research. Try to create your own self-folding shapes using the polystyrene. Using the marker, add dark lines to the areas you want to bend while heating in the oven. The ink from the marker causes the plastic to heat at a different rate than the plastic by itself. Turn on the oven light and watch while the plastic is heating to see the folding in action.*

ROTATIONAL RACES

LEVEL OF DIFFICULTY: MEDIUM

FROM BEGINNING TO END: 45 MINUTES

OTHER CATEGORIES: ENGINEERING, MATH

THE REAL QUESTION:

The **moment of inertia** is a rotating object's resistance to any change in its motion. How does the distribution of mass on an object impact its rotational speed? Form a hypothesis about how the distance of marbles to the axle of a rolling vehicle will impact how fast it rolls.

MATERIALS:

- Unsharpened pencil or wooden dowel
- Clay (or adhesive clay)
- 2 CDs
- 8 marbles
- Ruler
- Piece of wood or cardboard, about 1 foot (≈30 cm) by 4 feet (≈1.2 m)
- Protractor
- Stopwatch

THE STEPS:

1. The pencil or wooden dowel is the axle for your rolling racer. Build your rolling racer by using clay to attach the axle to the two CDs. The CDs will be the wheels.

2. Roll eight clay balls of equal size, about the size of the marbles. This clay will be used to attach the marbles to the wheels.

3. Space out four marbles like a clock at the 12, 3, 6, and 9 o'clock positions (like a plus sign) on the wheel. Attach the marbles in the clock locations close to the edge of the wheel using one clay ball for each marble. Do the same thing on the other wheel.

4. Measure and record the distance of the marbles to the axle. This distance should be the same for all the marbles.

5. Use the wood or cardboard to make a ramp. Using a protractor, set up the ramp so it has a 15- to 30-degree incline.

6. Place the racer at the top of the incline. Then let it roll down the ramp. Time how long it takes the racer to roll from the top to bottom of the ramp.

7. Repeat Steps 3 to 6, but each time move the clay closer to the axle. Put the marbles at least three different distances from the axle.

OBSERVATIONS:

- **How does the distance of the marbles to the axle of your vehicle affect how fast it rolls?**

CONTINUED ➔

The Real Science Behind How and Why: You can feel the impact of moment of inertia by sitting in a wheeled office chair with your feet up and arms stretched out wide. Have someone give you a little spin. As you spin, pull your arms close to your body. When you do this, you will spin faster. When something rotates (spins), it has a type of momentum called **angular momentum**.

Momentum is conserved. Pulling your arms in decreases your moment of inertia and increases your speed. It also works in reverse! If you put your arms back out, you will slow down. You experience the moment of inertia when you pull your arms in *and* when you stretch them out, because both times you are changing the rotational speed.

Now Try This! *Try adding more marbles or putting the marbles in different patterns. How can you make the rolling racer faster? Make two rolling racers and race them!*

RUBBER BAND POWER 2-IN-1

LEVEL OF DIFFICULTY: MEDIUM

FROM BEGINNING TO END:
20 MINUTES EACH

OTHER CATEGORIES: ENGINEERING, ART

THE REAL QUESTION:

How can a rubber band create motion? Form a hypothesis about how the amount of winding of a rubber band will impact motion. Learn about **potential energy** and **kinetic energy** by creating two different devices using similar materials. Start with steps 1–3, then decide if you would like to make a racing device, a drawing device, or both.

MATERIALS:

- **Toothpick or paper clip**
- **2 to 5 rubber bands that are about 2 inch (≈5 cm) x ¹⁄₁₆ inch (≈2 mm) in size**
- **2 metal washers**
- **Masking tape**
- **Skinny marker**
- **Clip-type clothespin**
- **Paper**
- **Spool of thread**

STEPS:

1. Use a toothpick or paper clip to push a rubber band through the hole of the spool so the rubber band is hanging out both ends of the spool.

2. On one end of the spool, pull the rubber band through the hole of a washer. Break a toothpick so it is smaller than the diameter of the spool. Secure the rubber band to one end of spool by looping it over the toothpick and taping the toothpick/looped rubber band to the spool end.

3. On the other end of the spool, again pull the rubber band through the hole of a washer.

To Make the Racing Device (Option A)

4. Stick the skinny marker into the rubber band loop next to the washer.

5. Hold the spool in one hand and twirl the marker a couple of times to wind the rubber band.

6. Place the spool sideways on a flat, smooth surface. The marker should be parallel to the surface. Let it go and watch it move. If the spool does not move well, trying winding the rubber band more or less.

CONTINUED ➜

To Make the Drawing Device (Option B)

4. Stick the clothespin into the rubber band loop next to the washer.

5. Clip the clothespin onto the marker. Lay a piece of paper flat. Adjust the marker so that the marker tip will touch the paper when the spool is in motion.

6. Hold the spool in one hand and twirl the clothespin a couple of times to wind the rubber band.

7. Take off the marker cap. Put the spool on the paper so that the marker tip is touching the paper and let go.

OBSERVATIONS:

➲ **How does the number of times you wind the marker impact the motion? How does the motion of the two devices compare?**

The Real Science Behind How and Why: Potential energy is energy that is stored. Potential energy exists in food or batteries (chemical energy), high places (gravitational energy), or in this case, the stretching of elastic rubber bands (mechanical energy). Energy can be converted from one form to another. Potential energy becomes kinetic energy once it is in motion.

Now Try This! *Try testing different spool sizes and different types of rubber bands. How can you modify your design to make it travel faster or straighter?*

DISSECTING SOUND

LEVEL OF DIFFICULTY: EASY

FROM BEGINNING TO END: 30 MINUTES

OTHER CATEGORIES: TECHNOLOGY, ART

THE REAL QUESTION:

How can you separate a mixture of sounds into separate frequencies? Form a hypothesis about how pitch relates to tube length. Learn about connections between pitch and frequency of sound. Build a device that can be used for eavesdropping or making music.

> **HIGHER PITCH OF SOUND =
> HIGHER FREQUENCY**

! **CAUTION:** Listening to loud noises can damage hearing. Have an adult help cut the PVC pipe if you use this material instead of cardboard paper towel rolls.

MATERIALS:

- 8 cardboard paper towel rolls or 90 inches (2.3 m) of 1-inch (≈2.5-cm) diameter PVC pipe, and a tool to cut PVC pipe into 8 pieces
- Ruler
- Scissors
- Masking tape
- String

THE STEPS:

1. Plan 5 different lengths to create 5 tubes of varying and increasing lengths. For example, if you are using 11-inch (≈28-cm) paper towel rolls, you can create lengths of 5½ inches (≈14 cm) (half a roll), 11 inches (28 cm) (one roll), 16½ inches (≈42 cm) (one and a half rolls), 22 inches (≈56 cm) (two rolls), and 27½ inches (≈70-cm) (two and a half rolls).

2. Measure, cut out, and tape as needed to create the five tubes.

3. Use tape and string to tie the tubes together in descending order by size as shown in the photo on page 37.

4. Thread one piece of string all the way through the longest tube. Keep it long enough so it can tie above the device and be used as a handle. Cut off the excess string.

5. Find a room with different background noises such as several people talking, music playing, and a television with the volume on. Listen to the noise through each tube.

OBSERVATIONS:

- How does the sound compare when listening through each tube? How does the sound change when you put the end of the tube completely against your ear?

CONTINUED →

The Real Science Behind How and Why:
The longest tube has a longer column of air inside it. It vibrates more slowly which creates a lower-frequency sound wave. This lower frequency relates to the lower-pitched sounds you hear in the longest tube. The shortest tube has a shorter column of air that vibrates more quickly. This higher frequency has a higher-pitched sound in the shortest tube. The same relationship exists if you turn the device into a xylophone. Longer tubes play lower notes, and shorter tubes play higher notes.

Now Try This! *How can the same materials be used to create a xylophone? How do the notes struck with a mallet constructed from a pencil versus a silly straw differ from each other? What happens to the sound when you add a ball of clay to the head of the pencil and silly straw mallet? What other objects can you find to create a xylophone or mallet?*

THE ORIGINAL FIDGET SPINNER

LEVEL OF DIFFICULTY: EASY

FROM BEGINNING TO END: 10 MINUTES

OTHER CATEGORIES: ENGINEERING

THE REAL QUESTION:

How do forces influence circular motion? Learn about **centripetal force** by building your own spinning device. Form a hypothesis about how the amount of clay affects the ease of spinning your device.

! CAUTION: Have an adult help break or cut the wooden sticks to size.

MATERIALS:

- Masking tape
- 2 wooden sticks (bamboo skewer, wooden dowel, or unsharpened pencil; you'll need a 6-inch (≈15-cm) piece and a 3-inch (≈7.5-cm) piece
- Clay

THE STEPS:

1. Tape the two wooden sticks together into an X shape. The two sticks should cross so that three sides of the X shape are the same length from the center (about 1½ inches [≈4cm]) and one side is longer (about 4½ inches [≈11cm]).

2. Roll a clay ball about 1 inch (≈2.5 cm) in diameter and stick it onto the end of the longest stick.

3. Place your pointer finger inside the X-shape on the side with the weight and spin. It takes some practice, so keep trying until you figure it out.

4. Test out different amounts of mass for the clay ball by adding or removing clay.

OBSERVATIONS:

- How long can you keep the device spinning? How does changing the length of the sticks or the angle between the sticks impact the motion? Which design is easiest to spin?

The Real Science Behind How and Why: Due to **inertia**, an object in motion continues to travel in a straight line until an outside force causes it to speed up, slow down, or change direction. Centripetal force is a force needed to make something change direction. In this case, the direction it is moving in is a circle. Here's another example: If you swing a cup around your head with a string, there is tension on the string. This tension is the centripetal force. If the string breaks or you let go, you take away the centripetal force, and the cup moves away in a straight line.

Now Try This! *Gently push a marble into the top of a packaged gelatin cup. Tie a string to the cup and reinforce with tape. Swing the cup around in a circle. What happens to the marble? Try this outside: Tie a string to an open plastic cup of water and try spinning it over your head. Both these activities relate to centripetal force.*

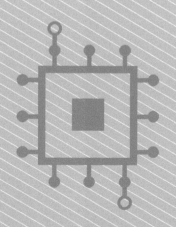

Chapter 3

Technology

The word technology is used in different ways. Technology can be *things* made for a purpose, such as cell phones or ink in ballpoint pens. Technology can also be *ways of making things*, such as how to make solar cells or bubble gum. Technology can even be *ways of doing things*, such as how to perform brain surgery.

In this chapter, you will investigate both the products and processes of technology. You will learn more about the science of sound and types of electromagnetic radiation, including visible light, radio waves, and ultraviolet (UV) light. You will also make your own technology, including a microscope, stethoscope, and motor.

For some of the experiments, you need items that are best borrowed from an adult. For Secret Patterns of Sound Waves (page 45) and for Light versus Sound (page 48), you will need a small speaker that can fit into a bowl. These experiments are easier with a wireless speaker, but a wired speaker will work, too. For Cell Phone Signal Blocking (page 59), you need a cell phone, and it's easier if you have two phones. Make sure

you have permission to borrow these materials and that you handle them with care.

You will need some electronic supplies that can be found at a hardware or hobby store. For the Homopolar Motor Color Whirl experiment (page 61), you need at least five inches of wire (16- or 18-gauge copper or electrical wire), one or more neodymium magnets (about 12 mm x 6 mm in size), and some AA batteries. You should also pick up a laser pointer (or two) from a dollar store (green or blue in color, if possible). For the Light versus Sound experiment, a small mirror with a diameter about 1 inch (≈2.5 cm) or smaller can be found at a dollar store or craft store.

Deeper understanding of science leads to advances in technology. Science has inspired many technologies, like new ways to grow food, or biodegradable plastic for your to-go cup, and nanomedicine for cancer treatments. This chapter is all about understanding science and technology. This is one step on the way to discovering the next innovative technological breakthrough!

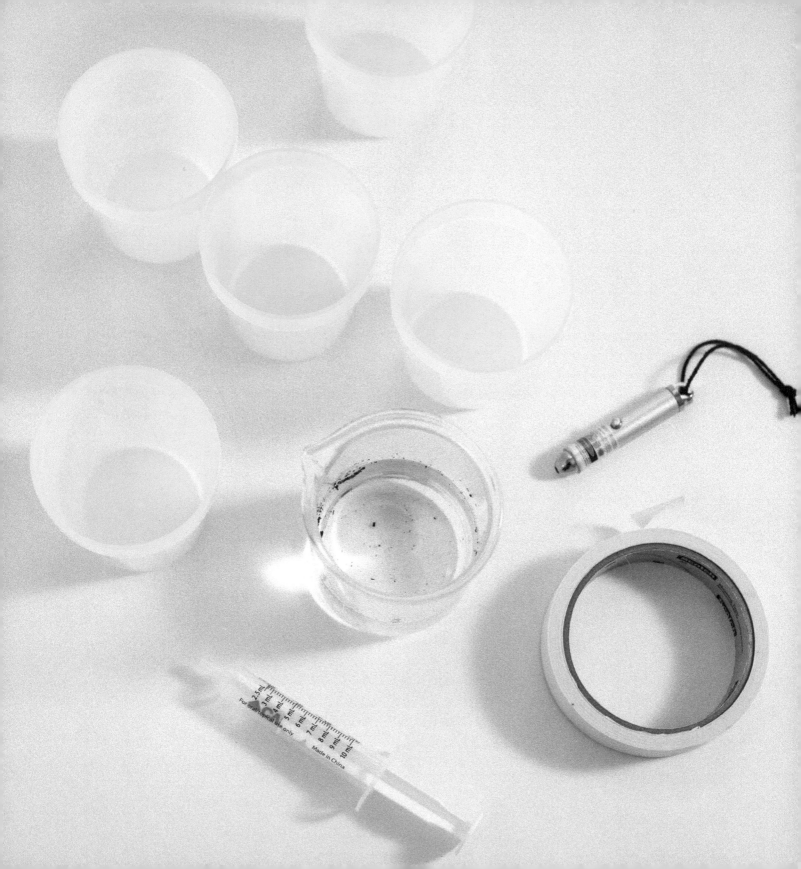

MAKE A MICROSCOPE

LEVEL OF DIFFICULTY: MEDIUM

FROM BEGINNING TO END: 30 MINUTES

OTHER CATEGORIES: SCIENCE, ENGINEERING

THE REAL QUESTION:

What can you observe about microscopic organisms in a water sample? Form a hypothesis about what you expect to see in a single drop of water. Create a microscope to see microorganisms in a water sample.

! CAUTION: Lasers can cause eye damage. Do not shine a laser pointer into someone's eye. Collect outdoor water samples with an adult.

MATERIALS:

- **5 plastic cups**
- **Masking tape or transparent tape**
- **Plastic syringe (oral medical syringe or cooking flavor injector syringe, without a needle)**
- **Water sample from a puddle, stream, or base of houseplant**
- **Laser pointer (green or blue if possible)**

THE STEPS:

1. Find a place to set up your microscope. A floor or table surface near a wall will work well.

2. Create a tower from two plastic cups. Place the rim of one cup down on a flat surface (floor or table) with the base facing up. Place the base of a second cup on top of the base of the first cup. Add tape where the two bases meet to attach the two cups. Repeat using two more cups to create a second tower. These two towers will be used to hold the plastic syringe.

3. Draw your water sample into the syringe.

4. Place the two cup towers closely together and rest the top of the syringe plunger on the rims. Add tape if needed to secure the plunger in place.

5. Use the remaining plastic cup with the rim facing down as a mount for the laser pointer. Rest the laser point on the base of the cup. Tape down the button of the laser pointer if you do not want to have to hold the button to keep the laser pointer turned on.

6. Slowly press the plunger on the syringe to form the largest water droplet you can without it dripping.

CONTINUED ➜

7. Move the laser light on the base of the cup until it lines up directly with the water drop. Run the beam of light down the center of the syringe. Either rest your hand on the base of the cup to hold the laser beam in place or secure it with tape.

8. Observe the shape and motion of the microorganisms in your water sample. If your wall isn't white in color, it may help to tape a white sheet of paper to the area where the laser projects the image.

OBSERVATIONS:

◉ **Record the shape and motion of the microorganisms you see in your science notebook.**

The Real Science Behind How and Why:
The water droplet hanging from the syringe is used as a spherical lens. Refraction of the laser light beam causes a magnified image to appear on the wall. Common micro-organisms found in puddles and ponds include *Daphniidae* (water fleas), *Culicidae* (mosquito) larvae, and **protists** such as paramecia or amoebas.

Now Try This! *Try out different distances between the laser, water drop, and wall to find the ratio that creates the largest, clearest image. Compare different colors of laser light to see which color creates the clearest image.*

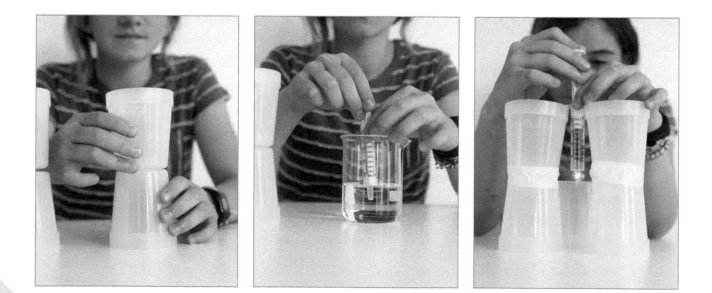

SECRET PATTERNS OF SOUND WAVES

LEVEL OF DIFFICULTY: EASY

FROM BEGINNING TO END: 15 MINUTES

OTHER CATEGORIES: SCIENCE, ENGINEERING

THE REAL QUESTION:

How does the volume and tone of sound affect the motion of salt or sugar? Form a hypothesis about how you think the salt or sugar will move and why. Learn about sound waves and frequency with this feat of **acoustical engineering**.

! CAUTION: Playing sounds at loud volumes may damage your ears or speakers. Use a safe volume, and wear earplugs if you are sensitive to sound.

MATERIALS:

- Portable wireless speaker
- Glass bowl (large enough for the speaker to fit inside)
- Cell phone paired to the wireless speaker
- Plastic wrap
- Masking tape
- Salt or sugar

Note: If you do not have a portable wireless speaker, you can modify this experiment. Alternative materials include speakers that can play music, a small cardboard box, and scissors. Reusing an empty box of fruit snacks or granola bars for the small cardboard box works well for this experiment. Cut the box in half along its length to create a "cookie sheet" piece of cardboard. Place this piece of cardboard on the speaker, sprinkle salt or sugar in it, and finish with steps 5 and 6 of the experiment.

THE STEPS:

1. Put a portable wireless speaker into a large glass bowl. Turn the speaker on and pair it with the cell phone.

2. Cover the top of the bowl with plastic wrap. Make sure to stretch the plastic wrap so it is tight and flat across the bowl's top.

3. Place a ring of tape standing up along the outer edge of the bowl. Do not fold the tape down because it is a barrier to ensure that the salt does not fall over the edge of the bowl.

4. Sprinkle a small amount of salt or sugar over the plastic wrap on the bowl.

CONTINUED →

5. Play a song or sound sample through the speaker. Start with a low volume and slowly increase the volume.

6. Add more salt or sugar as needed.

OBSERVATIONS:

● How does the motion of the salt or sugar change as the volume increases? How do the patterns created change with different tones?

The Real Science Behind How and Why: Sound can move through gases, liquids, and solids. Sound is a wave made by the back-and-forth vibration of particles. Frequency is a measure of how fast a particle vibrates back and forth. As a sound wave vibrates faster, its frequency increases. We hear this as a higher pitch. The unit used for frequency is **hertz**, often abbreviated to "Hz." Ernst Chladni studied sound during the 1800s in a way similar to the salt and sound experiment you just did. Instead of a speaker, he used a violin bow and moved it against a metal plate covered with sand. Research "Chladni plate" or see the Smithsonian's article on Chladni Plates in the References section (page 135) to see more of his work in action today.

Now Try This! *Use a free tone-generator website or a phone app to play specific frequencies of sound to the speaker. How does a lower frequency compare to a higher frequency? Replace the glass bowl with other household containers such as a plastic bowl or metal pot. How do different bowl materials and bowl sizes affect the results?*

LIGHT VERSUS SOUND

LEVEL OF DIFFICULTY: MEDIUM

FROM BEGINNING TO END: 30 MINUTES

OTHER CATEGORIES: SCIENCE, ENGINEERING

THE REAL QUESTION:

How does the volume and tone of sound affect the changing vibrations of a mirror? Form a hypothesis about how the vibrations of sound on a mirror influence the path of a laser light beam. Learn about incident angles and reflected angles by creating a waveform visualizer.

! CAUTION: Keep the beam of light of the laser pointer away from people's faces. Never look directly at the laser pointer light. Handle your mirror carefully to avoid getting cut.

MATERIALS:

- Sound setup from the Secret Patterns of Sound Waves experiment on page 45 (bowl, plastic wrap, and speaker)
- Glue
- Small mirror, about 1½ inches (≈4 cm) in diameter or smaller
- 4 clip-type clothespins
- Masking tape or transparent tape
- Laser pointer

*If you are using a dental mirror, pry the mirror from its stem using tweezers.

THE STEPS:

1. Use your bowl, plastic wrap, and speaker setup from the Secret Pattern of Sound Waves experiment (page 45).

2. Glue a small mirror to the surface of the plastic wrap reflective-side up.

3. Create a stand for the laser pointer. Clip three clothespins to the bottom of the pencil to create a base. Add tape for stability. Tape down the button on the laser pointer so it is always on. Then, tape the laser to the fourth clothespin and clip to the top of the pencil.

4. Adjust the laser beam so it points directly onto the mirror. If you turn your bowl sideways, you can reflect the beam onto the wall. Otherwise, the beam will reflect onto the ceiling.

5. Play a song or sound sample through the speaker. Start with a low volume and slowly turn it up. Enjoy your laser light show!

OBSERVATIONS:

- **What impact do changes in volume or tone have on the light show? How does changing the angle of the laser affect the light show?**

The Real Science Behind How and Why:
The angle that the beam of light from the laser enters the mirror is called the incident angle. The angle that the light exits the mirror is called the reflected angle. The vibrations from the speaker's sound also cause the mirror to vibrate. The Law of Reflection says that the angle of incidence is equal to the angle of reflection. Every vibration slightly changes the incident angle of laser light, which then changes the reflected angle. Light travels very fast, so the changing vibration is shown by the changing patterns of the laser light. Although this is not what happened in this experiment, it is possible to turn sound into light. Research the sonoluminescence of mantis shrimp to learn more!

Now Try This! *Use a cell phone to take a time-lapse video of the laser motions. Try using different types of flashlights instead of the laser pointer to compare how different beams of light react.*

HYDROPONIC SYSTEM

LEVEL OF DIFFICULTY: EASY

FROM BEGINNING TO END: 1 WEEK

OTHER CATEGORIES: SCIENCE, MATH

THE REAL QUESTION:

How does a seed grow? Hydroponics is a method of growing plants without soil. Form a hypothesis about how quickly a plant will grow using hydroponics. Learn about plants by building and using your own hydroponic growing system.

MATERIALS:

- 2-liter soda bottle
- Scissors
- Water
- 6 to 8 inches (≈15 to 20 cm) thick cotton string or twine
- Growing material (cotton balls or piece of cloth like a clean cotton sock)
- Seeds (green bean, mung bean, tomato, lettuce, spinach, arugula, or basil work best)

THE STEPS:

1. Cut the soda bottle in half right where it starts to curve at the top. Flip the top half of the bottle upside down and put it inside the bottom piece so it looks like a funnel.

2. Fill the bottom bottle section about ¼ of the way with water.

3. Thread the string through the neck of the bottle. The string needs to rest on the bottom of the bottle. Wrap the rest of the other end of string around the bottom of the top funnel piece.

4. Pull and stretch your growing material until it fills the inside of the top "funnel" section.

5. Place 2 to 5 seeds in the center of the growing material.

6. Dampen the growing material with a small amount of water. The string will continue to provide water during the hydroponic gardening process.

7. Place the bottle in a sunny spot and watch your seed grow. Add water to the bottom bottle as needed over the next week.

CONTINUED →

OBSERVATIONS:

- Check your plant daily and record any changes in your science notebook.

The Real Science Behind How and Why:
The hydroponic system you created is called a wick system. The wick draws the nutrient solution from a reservoir up into the growing material. However, the most commonly used hydroponic system is a drip system in which a timer and pump are used to drip nutrient solution onto the growing material. Hydroponic crops can grow larger, quicker, and in less space than in a traditional soil field. They can grow indoors or even in space, as they have aboard the International Space Station. Agricultural engineers design hydroponic growing systems so the perfect amount of water, nutrients, and oxygen are always available for plants.

Now Try This! *Your plant will sprout but cannot continue to survive on water alone. Try using water from a fish tank or old tea bags to make a nutrient solution for the plants. Build additional hydroponic systems, and test out which solution works best.*

MAKE A KNOCKOFF ULTRAVIOLET LIGHT

LEVEL OF DIFFICULTY: EASY

FROM BEGINNING TO END: 15 MINUTES

OTHER CATEGORIES: SCIENCE

THE REAL QUESTION:

Ultraviolet (UV) light is a type of electromagnetic radiation that humans cannot see. It has a shorter wavelength and higher frequency than visible light. What can be seen with UV-like light? Form a hypothesis about which items will glow under a UV-like light and why. Learn about **fluorescence** by making your own UV light.

MATERIALS:

- Flashlight
- Clear tape or plastic wrap
- Blue permanent marker
- Purple permanent marker
- Yellow highlighter marker
- Paper

THE STEPS:

1. Cover the lens of the flashlight with clear tape or plastic wrap. Color it with the blue permanent marker.

2. Add another layer of clear tape or plastic wrap. This time color it with the purple permanent marker.

3. Add a third layer of clear tape or plastic wrap. Color it with the blue permanent marker.

4. Draw with the yellow highlighter on paper and test it under your light. Find a dark place and turn on the flashlight.

OBSERVATIONS:

- Make a list of all the items that glow under your light in your science notebook. How do the items that glow differ from the items that do not glow?

CONTINUED →

The Real Science Behind How and Why:
Things invisible to the human eye become visible under the rays of ultraviolet (UV) light. Fluorescent molecules absorb the short wavelengths of UV light and instantly reflect it back. Biological stains from saliva, blood, and urine will glow under UV light. The light created in this experiment is similar to light from an incandescent black light bulb. This type of blue light is part of the visible light spectrum which can make things glow a little, but is not technically ultraviolet light.

The technology of using ultraviolet light is used for more than just analyzing stains. UV light can help determine if paintings and signatures are real or replicas and can illuminate unseen fingerprints or ink remnants.

Now Try This! *Soak a yellow highlighter in a small amount of warm water for a couple hours to create a solution that glows under UV light. Now use the highlighter solution instead of water in recipes to make your own glowing bubbles or slime. Which recipe glows most brightly?*

SUPER SIMPLE STETHOSCOPES

LEVEL OF DIFFICULTY: EASY

FROM BEGINNING TO END: 15 MINUTES

OTHER CATEGORIES: SCIENCE, **MATH**

THE REAL QUESTION:

How can you observe your heartbeat using sound or sight? Form a hypothesis about how a method using sound compares to a method using sight for measuring your **heart rate**. Learn about how the stethoscope was invented by building your own.

MATERIALS:

- Paper
- Mini marshmallow
- Toothpick
- Stopwatch

THE STEPS:

1. Paper Method: According to the National Center for Biotechnology Information, Dr. René Laënnec got his inspiration for making a stethoscope after he saw two children sending signals to one other. The first child was scratching one end of a long piece of wood with a pin. The other child was putting their ear on the other end, listening to the amplified sound of the scratching pin. This made Dr. Laënnec think about how sound travels. One day in 1816 he rolled a sheet of paper into a tube. He placed one end over the chest of his patient and the other end to his ear. Go ahead and try it for yourself!

2. Mini Marshmallow Method: Poke a toothpick into the top of a mini marshmallow, just deep enough so it stands up. Lay your arm flat with your palm facing up. Look at your wrist and place the marshmallow above your vein. The toothpick should twitch each time your heart beats.

3. Use a stopwatch to count how many times your heart beats in one minute.

OBSERVATIONS:

- How many times does your heart beat in one minute? How does your heart rate measurement using the paper compare to using the marshmallow? What could be the sources of error between these two methods?

The Real Science Behind How and Why: The invention of the stethoscope is a great example of design thinking. Before stethoscopes were invented, a doctor would put his ear directly on a patient's chest to listen to the person's heartbeat. This could be an uncomfortable violation of the person's personal space, so Dr. René Laënnec wanted to do something different. Design thinking is a problem-solving process that focuses on caring about people. The first step of design thinking is to empathize, which means that you think about and understand how others feel. Next, you define your problem. Then you brainstorm as many ideas as you can. Finally, you prototype and test your best ideas. It is a lot like the scientific method! How can you use design thinking to solve a problem in your life?

Now Try This! *How do different activities change your heart rate? Measure your heart rate when you first wake up and after you exercise.*

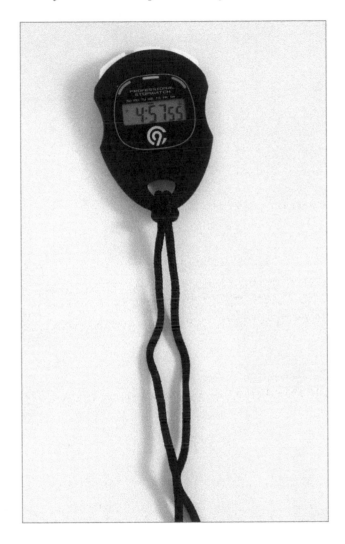

CELL PHONE SIGNAL BLOCKING

LEVEL OF DIFFICULTY: MEDIUM

FROM BEGINNING TO END: 30 MINUTES

OTHER CATEGORIES: SCIENCE, ENGINEERING

THE REAL QUESTION:

How can you prevent reception to a cell phone or a radio? Form a hypothesis about how much aluminum foil is needed to block radio waves. Build your own **Faraday cage** and learn more about the electromagnetic spectrum.

! CAUTION: Electronic devices like cell phones are expensive. Handle with care.

MATERIALS:

- 2 cell phones or battery-powered radios
- Aluminum pan or cardboard box covered in aluminum foil, 9 inches (≈23 cm) by 13 inches (≈33 cm)
- Aluminum foil
- Ruler
- Scissors
- Masking tape or transparent tape

THE STEPS:

1. Make a test call to make sure the phone works and the volume is up.

2. Put the phone into the pan and call it again to make sure that it rings. If it does not ring, the pan may be too deep. Put something under the phone to raise it closer to the top until it does ring.

3. Cover the pan with a sheet of aluminum foil and turn down the edges. Test to see if the phone rings inside the aluminum pan with the aluminum foil cover. It should not ring. If it does, make sure the cover is wrapped tightly around the top of the pan and try again.

4. Take off the aluminum foil cover. Cut ½-inch (≈13-mm) wide strips of foil that are a bit longer than the longest side of your aluminum pan. You will need 15–20 strips of foil.

5. Arrange the foil strips in a grid pattern across the top of the aluminum pan. Try calling the phone after each new grid pattern. Test out different spacings to figure out which size hole blocks the signal. Place the phone so you can see the display showing its signal strength.

CONTINUED ➜

OBSERVATIONS:

⊕ **Which size holes blocked the signal? What happens if you arrange the foil strips all in one direction instead of a grid pattern?**

The Real Science Behind How and Why: Radio waves are a very important type of electromagnetic radiation that carries information for TV, radio, wireless internet, and cell phones. Radio waves have the longest wavelength of any type of electromagnetic radiation, ranging in size from 1 mm to 10,000 km. A Faraday cage is made from conductive materials that shield against radio frequencies or electric and magnetic fields (EMFs). Faraday cages are often discussed in reference to protecting against an electromagnetic pulse (EMP), which can ruin electronic devices. Research electromagnetic pulses to learn more about their natural and human-made causes.

Now Try This! *What other objects and materials can you find that work as a Faraday cage?*

HOMOPOLAR MOTOR COLOR WHIRL

LEVEL OF DIFFICULTY: EASY

FROM BEGINNING TO END: 30 MINUTES

OTHER CATEGORIES: SCIENCE, ENGINEERING, ART

THE REAL QUESTION:

How do different patterns of colors appear to change when spinning? Form a hypothesis about how colors will appear in a spinning wheel. Build a homopolar motor to investigate colors and vision.

! CAUTION: Complete this experiment with an adult. When the circuit is connected, the battery will get hot. Do not let small children or animals play with the magnets. Do not place magnets near any electronics equipment such as tablets or phones. Store the magnets in a safe place when not in use.

MATERIALS:

- ⊘ AA battery
- ⊘ Neodymium rare earth magnet (about 12 mm x 6 mm)
- ⊘ Steel wood screw
- ⊘ 16- or 18-gauge copper or electrical wire, about 5 inches (≈13 cm) in length
- ⊘ Cup or lid about 3 inches (≈8 cm) in diameter for circle template
- ⊘ Computer paper
- ⊘ Markers (red, blue, yellow, and other colors)
- ⊕ Scissors

THE STEPS:

1. Assemble the homopolar motor. Hold the AA battery with the negative (flat) end facing up. Attach the magnet on the flat side of the screw. Touch the pointed end of the screw to the positive (pointed) end of the battery. With your free hand, take one end of the wire and touch it to the magnet hanging from the screw. The nail and magnet should spin.

2. Trace three circles onto the computer paper with a circle template and a marker. Cut out the circles.

CONTINUED →

3. Color in the circles with three different patterns. For the first pattern, divide one circle into three approximately equal-sized sections, like three pieces of pie, and color one section red, one blue, and one yellow. Choose the number, size, and color of pie pieces for the two remaining circles.

4. Place one of the colored paper circles, centered and colored side up, to the magnet attached to the metal screw head. Place the metal washer on the center bottom side of the colored paper circle. Magnetism will hold the colored circle in place.

5. Reassemble your homopolar motor, now with the added colored paper circle and metal washer. Again, take one end of the wire and touch it to the magnet hanging from the screw. This time, the nail, magnet, and colored paper circle will all spin.

OBSERVATIONS:

● **How do the colors on the paper circles appear to change when spinning?**

The Real Science Behind How and Why:
A homopolar motor is an electric motor that uses direct current from a battery to power rotational movement. For this experiment, this rotational movement is used to spin a colored disc. The primary colors of light are red, blue, and green, and equal amounts of each of these three colors of light combine to form white light. The retina in the eye contains three types of light-sensing cells shaped like little cones: red-detecting, green-detecting, and blue-detecting. These are the only colors of light that our eyes can detect. It is different proportions of just these three colors that make all the other colors we can see! This is called additive light theory and is not the same as mixing pigments like paint colors, where the primary colors are red, blue, and yellow. Research to learn more about the similarities and differences between additive and subtractive color theory.

Now Try This! *Instead of pie pieces, create different patterns of colors and investigate how it impacts the results. Find a different way to spin the colored circle, like using a pencil pushed through the center of the circle to spin it like a top. How many different methods can you find?*

Chapter 4

Engineering

Engineering is all about applying science and math knowledge to help solve problems and reach goals. These problems and goals are important for improving life on our planet. A group of experts from around the world came together as part of the National Academy of Engineering (NAE) to make a list of what they called "the Grand Challenges for Engineering in the 21st century." This list includes important life-changing goals like making solar power better. Right now only about 1 percent of the world's power comes from solar energy, but this number has the potential to be much higher. The sun provides more energy in just one hour than people worldwide use in a year!

The experiments in this chapter are a step in the direction of helping to engineer a better future. You will construct better buildings using **bioglues**, create a candle-powered boat, and even design a rock-hopping robot! All these experiments follow the engineering design process in which you ask questions, imagine solutions, make a plan, create your idea, and improve your design through creative problem solving. The engineering design process is a lot like the scientific method, with one exception. The scientific method and the engineering design process have two different goals. The goal of the scientific method is to figure out how the natural world works. The goal of the engineering design process is to solve a problem using limited materials and a budget. The experiments in this chapter all focus on solving a problem.

You probably have everything at your house to complete these experiments. For the Make Your Own Molds experiment (page 69), you will want to check your kitchen for food-safe glycerin and unflavored gelatin. If you don't have these items, you should add them to the shopping list for your next trip to the grocery store.

BUILDING WITH BIOGLUE

LEVEL OF DIFFICULTY: MEDIUM

FROM BEGINNING TO END: 45 MINUTES

OTHER CATEGORIES: SCIENCE

THE REAL QUESTION:

Bioglue is a glue made from natural materials like sugar. Which bioglue creates the strongest structure? Engineer a building, test it to its destruction, and maybe even make a tasty treat! Form a hypothesis about the building design and glue that you think will be best and why.

⚠ CAUTION: Get help from an adult if you melt any items using a stove or microwave. Use caution when handling heated materials. If you have food allergies, avoid using those foods. This experiment may be messy.

MATERIALS:

- ➲ Disposable tablecloth or newspapers
- ➲ Several sugar-based materials, such as frosting, marshmallow cream, honey, gummy bears, marshmallows, and caramel
- ➲ Graham crackers
- ➲ Pile of books (for weight)
- ➲ Spoon, butter knife, or craft stick to stir and spread bioglue

THE STEPS:

1. Cover your workspace with a disposable tablecloth or newspapers to prevent messes. Create two (or three) different bioglues to test, using several of the sugary materials on the list. Experiment with mixing and melting your own combination of materials like honey, gummy bears, marshmallows, or caramel.

2. Decide on a design to use for each of your graham cracker buildings. You can design a building to have four walls to create a rectangular or square base, or three walls to create a triangular base. Use the width or length of an entire graham cracker–or two– to be the size of each side of the building.

CONTINUED →

3. Each building must have the same size and shape but will use different glue. Make sure your design is wide enough so that you can stack books on the building.

4. Put together your graham cracker buildings, each with its special bioglue. Depending on your bioglue, it may help to wait for the glue to harden before adding the next piece. Try refrigerating sections of your graham cracker buildings before attaching them.

5. Once your buildings are finished, have a friend or family member add one book at a time onto each building until it breaks. Use your hands to hold the books steady. Be sure to use the same books for each building. Record your results in your science notebook.

OBSERVATIONS:

⊙ **How many books did each building support? What was the weakness that caused each building to break?**

The Real Science Behind How and Why: We make glues for many different applications, from the temporary hold of a sticky note to the strength of superglue. Nature also makes its own glues. Scientists study the toe pads of geckos and the cement made by ocean clams to learn more about glue. **Biomimicry** is a fancy word to describe new technologies made using inspiration from nature. For example, scientists at North-western University studied slug slime to make a stretchy, nontoxic bioglue that can even stick to wet surfaces. Bioglues like this are helpful during surgery or when joining surfaces underwater.

Now Try This! *Create a new building design or bioglue recipe to test out. Use a scale to find how much weight the building can support.*

MAKE YOUR OWN MOLDS

LEVEL OF DIFFICULTY: MEDIUM/HARD

FROM BEGINNING TO END: 2 DAYS

OTHER CATEGORIES: TECHNOLOGY, MATH

THE REAL QUESTION:

How can you mass produce chocolate candy?
Form a hypothesis about how long it will take to
create five pieces of identical chocolate candies.
Learn about the manufacturing process from
conception to production.

> **!** CAUTION: Have an adult help use the
> oven. Handle hot items carefully. Not all
> glycerin is meant to be used with food, so
> check the label to make sure it is food safe. Do
> not eat the mold material. If you have allergies
> to any of the ingredients in chocolate chips,
> use water instead.

MATERIALS:

- 1 ounce box of unflavored gelatin
  (4 packets)
- ¾ cup (≈177 ml or 6 ounces) of food-grade
  glycerin
- Large mixing spoon
- Microwavable glass bowl, large enough to
  hold 4 cups (≈1 liter)

- ¾ cup (≈177 ml) boiling water
- Small cardboard box (slightly larger than
  the item you are molding)
- Aluminum foil
- Items to make mold (small, hard plastic toys
  that sink in water)
- Fine mesh strainer
- ½ cup of chocolate chips (≈3 ounces
  or 90 grams)

THE STEPS:

1. Read over the entire experiment and make
   a plan to streamline your chocolate-making
   process. Record your start time in your
   notebook.

2. Mix the gelatin and glycerin in the glass
   bowl. Allow the mixture to sit while you wait
   for water to boil.

3. Carefully add the boiling water and stir
   slowly. Do not let air bubbles get into your
   mixture!

4. Line the small cardboard box with aluminum
   foil. Make sure it is watertight.

CONTINUED →

5. Place the item you will use as a mold into the box. Stir the mold mixture again and then slowly pour it through the strainer into the aluminum foil. The mold must be completely covered. (You can repeat this process with another item and a box lined with aluminum foil if you have enough remaining mixture.)

6. Let the mixture sit at least 3 hours to overnight in the fridge so it can solidify. Once the mold is solid, carefully take it out of the box and peel away the aluminum foil. Pick up the mold and use your thumbs to slowly push your item out of the mold material. You are now ready to make some chocolate molds! If the aluminum foil will not peel off, place it under a stream of warm water in the sink and then try again.

7. Add the chocolate chips to a microwavable glass bowl and microwave for 30 seconds. Stir and then microwave for an additional 30 seconds. Continue heating and stirring for 15-second intervals until the chocolate is mostly melted. Let it sit, and the last few pieces will melt.

8. Carefully pour your chocolate into your mold. Put the mold into the refrigerator or freezer to cool. If you can't use chocolate, pour water into your mold and place into the freezer to make ice cubes shaped like your chosen item!

OBSERVATIONS:

⊛ **How long does it take to create five pieces of identical chocolate candies? How does putting the chocolate in the freezer compare to putting it in the refrigerator?**

The Real Science Behind How and Why: The original object from which a mold is made it called a model. The replica of the model made using a mold is called a cast. Many plastic (and even some metal) objects you interact with on a daily basis were cast in a process similar to this one. Just as in manufacturing, your mold will get damaged after many uses, and a new mold will need to be made. This is true for all mass-produced objects, such as action figures.

Now Try This! *You can re-melt the gelatin mixture in the microwave for about 1 to 4 minutes (test and stir as you go) and re-pour to make a new mold. Repeat all the steps to make changes and improve the process.*

WALK-ALONG SPINNING FISH

LEVEL OF DIFFICULTY: MEDIUM

FROM BEGINNING TO END: 15 MINUTES

OTHER CATEGORIES: SCIENCE, TECHNOLOGY

THE REAL QUESTION:

How can you control the movement of a spinning fish? Form a hypothesis about how paper size affects the rate of spin. Test a flying fish to practice the same skills used by aeronautical engineers.

MATERIALS:

- **Thin, lightweight paper, like a thin newspaper**
- **Ruler**
- **Scissors**
- **2-foot by 3-foot (≈0.6-m by 1-m) piece of cardboard or foam core board**

Note: This experiment is best done in a large indoor space with little air movement, such as a hallway or an empty gym.

THE STEPS:

1. Cut a strip of paper 8 inches (≈20 cm) long and 1 inch (≈2.5 cm) wide.

2. Create two cuts on the long side ½ inch (≈1–2 cm) from each end. Each cut needs to be on opposite sides that go halfway through the short width of the strip.

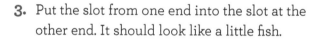

3. Put the slot from one end into the slot at the other end. It should look like a little fish.

4. Create additional "fish" of different dimensions (length, width, distance of cuts from end).

5. Pinch one side of the fish, hold it up high, and let it go. Watch it fly! Try holding your fish in different locations so it falls away from you instead of toward you when you let it go.

6. As your fish spins down and away, walk forward with your air pusher (piece of cardboard or foam core board) slanted in front of you at a 45 degree angle, so the top side is closer to you and the bottom side is farther away. Be sure to hold the air pusher this way the entire time you are walking so the air is pushed up. With a LOT of practice, you can keep your fish floating on a wave of air in front of you.

OBSERVATIONS:

- What happens if you change the length or width of the paper strip? What if you make the tails longer or shorter? Which design had the longest flight? Which design had the shortest flight?

The Real Science Behind How and Why: Air acts like water. Pushing your cardboard quickly in a pool of water would cause a fast-moving wave. Similarly, when you move your cardboard through the air, air rushes up and creates a wave of higher air pressure. The fish spins in front of the moving cardboard, riding on that wave of air. Research tumble-wing gliders to learn more about flying walk-along planes.

Now Try This! *Create flying fish using different types of paper, like a magazine cover or loose-leaf notebook paper. Which paper reaches the ground the fastest?*

PUT A CORK IN IT

LEVEL OF DIFFICULTY: EASY

FROM BEGINNING TO END: 30 MINUTES

OTHER CATEGORIES: SCIENCE, TECHNOLOGY

THE REAL QUESTION:

How can a chemical reaction power a rocket? Form a hypothesis about the amount of vinegar that results in the highest launch. Learn about equal and opposite reactions by launching vinegar and baking soda rockets.

CAUTION: Wear safety goggles and wash your skin if vinegar touches you. Never point the end of the cork rocket at any living thing or put your face in the path of the cork. If the bottle fails to launch, do not approach it for 5 minutes.

MATERIALS:

- **3 pieces of corrugated cardboard, about 6 inches (≈15 cm) by 12 inches (≈30 cm)**
- **Scissors**
- **Packing tape**
- **Empty and clean 20-ounce soda bottle**
- **Funnel**
- **1 gallon (≈3.8 liters) white vinegar**
- **Toilet paper**
- **Baking soda**
- **Wine cork or rubber stopper that can plug the soda bottle**

THE STEPS:

1. Trace and cut 3 or 4 rocket fins from the cardboard. Since these fins will get splashed with water and vinegar, cover them with packing tape to protect the cardboard.

2. Turn the soda bottle upside down and attach the rocket fins to it with tape, allowing the mouth of the bottle to stand about 1½ to 2½ inches (≈4 to 6 cm) above the ground when the rocket stands on the fins.

3. Use the funnel to fill the soda bottle one-third full of vinegar.

4. Using one square of toilet paper, place about 2 teaspoons (≈10 grams) of baking soda in the middle and roll it into a tube that can fit into the mouth of the bottle.

CONTINUED →

5. When ready to launch, quickly drop the toilet paper baking soda package into the bottle and use the wine cork to plug the bottle. Set the bottle on its fins and quickly move away. Watch the rocket as it launches to see how far it goes into the sky. Recording a video of each launch can make it easier to compare launch heights.

6. Repeat the steps to launch again, except fill your soda bottle one-half full of vinegar. Repeat a third time, making the bottle two-thirds full of vinegar.

OBSERVATIONS:

➔ **What happens when the toilet paper dissolves? How do the varying amounts of vinegar affect the launch?**

The Real Science Behind How and Why: The baking soda and vinegar mix create a chemical reaction that converts these materials into carbon dioxide (CO_2) and water (H_2O). As the CO_2 is produced, the pressure in the bottle increases until it pushes the cork and remaining water/vinegar mixture out the bottom of the bottle. Newton's Third Law is that *for every action there is an equal and opposite reaction*. When the water pushes out the bottom of the bottle, the bottle shoots up.

Now Try This! *What happens if you use the same amount of vinegar each time but different amounts of baking soda? What happens if you use 1-ply toilet paper instead of 2-ply?*

CANDLE-POWERED BOAT

LEVEL OF DIFFICULTY: MEDIUM

FROM BEGINNING TO END: 30 MINUTES

OTHER CATEGORIES: SCIENCE, TECHNOLOGY

THE REAL QUESTION:

How can the flame of a candle power a boat? Form a hypothesis about how changes in temperature can create motion. Learn about **convection currents** by building a boat powered by a candle's flame.

! **CAUTION: Have an adult help light the candle. Use caution around an open flame. Do not leave a pool of water unattended around pets or small children. Empty the pool of water when you finish the experiment.**

MATERIALS:

- **Aluminum foil, about 12 inches (≈30 cm) by 24 inches (≈60 cm)**
- **Scissors**
- **Transparent tape or paper clips**
- **Shallow pool of water, such as a bathtub or kiddie pool**
- **Cold water**
- **Tea light candle**
- **Lighter or match**

Note: This experiment works best indoors where there is little air motion.

THE STEPS:

1. Fold the aluminum foil to create the hull of the boat. Aim to create a hull that is about 2 inches (≈5 cm) wide by 2 inches (≈5 cm) tall by 4 inches (≈10 cm) long. It must be able to float while carrying the tea light candle. If it does not float, try again using different dimensions and thicknesses. Using one to two sheets of aluminum foil in thickness can create a boat that floats easily.

2. Cut out about 4 inches (≈10 cm) by 6 inches (≈15 cm) of aluminum foil to create a sail. Attach it to the hull of the boat using tape or paper clips. Shape the sail so it curves over the space above the candle's flame.

3. Fill a shallow pool with 2 inches (≈5 cm) of cold water. Let the water settle until it has no movement.

4. Place the tea light candle in the boat under the sail and place the boat in the water. Carefully light the candlewick. Observe the motion of the boat.

5. Blow out the candle and allow the boat to cool for one minute before touching it. Adjust and reshape the sail as desired and try again.

CONTINUED ➡

OBSERVATIONS:

◉ **Record your observations about the motion of the boat in your science notebook.**

The Real Science Behind How and Why:
The boat moves due to a convection current. The air around the candle is heated and therefore less dense. This rising warm air is directed away from the boat due to the sail. Colder, denser air fills its place. The candle then also heats this air, causing the process to continue. This creates a movement of air called wind that causes the boat to move. Although the amount of wind created in this experiment is very small, a similar process occurs in nature. In nature, wind is caused by the sun heating the Earth's surface unevenly, which sets convection currents in motion.

Now Try This! *Create sails of different shapes and sizes and compare how they affect the motion of the boat. How can you use a ruler and a stopwatch to determine the speed of your boat? (Hint: Speed is distance divided by time.)*

SPAGHETTI STRENGTH

LEVEL OF DIFFICULTY: MEDIUM

FROM BEGINNING TO END: 30 MINUTES

OTHER CATEGORIES: SCIENCE, ENGINEERING, MATH

THE REAL QUESTION:

How strong is a piece of spaghetti? Form a hypothesis about how much weight one to five pieces of spaghetti can support. Learn about strength of materials by building simple beam bridges.

! CAUTION: Wear protective glasses to shield your eyes against breaking spaghetti pieces.

MATERIALS:

- 2 identical chairs
- Uncooked spaghetti
- Paper clip (jumbo or #1 size)
- Sandwich-size plastic zip-top bag
- Masking tape or transparent tape
- Material to use as weights (coins, marbles, washers)

THE STEPS:

1. Move the base of two chairs together so there is a gap between to fit the length of one piece of spaghetti, leaving about 1 inch (≈2.5 cm) of space on both sides of where the spaghetti rests on the chair.

2. Open a paper clip such that it creates two hooks. Carefully shove one hook through both sides of the plastic bag in the center just under the zipper top.

3. Tape the ends of the spaghetti to each chair.

4. Hang the plastic bag on the center of the piece of spaghetti "bridge beam."

5. Hold the bag up and add one weight. Slowly lower the bag after adding each weight. Continue adding weights to the plastic bag one at time until the spaghetti breaks.

6. Record the number of items supported by one piece of spaghetti.

7. Repeat this process using bridges made of two, three, four, and five pieces of spaghetti. Tape the bundle of spaghetti together on each end before taping it to the chairs.

CONTINUED ➔

OBSERVATIONS:

- Create a graph showing the number of spaghetti pieces on the x-axis (horizontal) and number of items supported on the y-axis (vertical). What patterns do you notice in the data?

The Real Science Behind How and Why: With a single piece of spaghetti, all the force is applied to one brittle support. By bundling the spaghetti strands, you are creating something more like a rope or a cable. The load is shared by all the strands (the bottom strands more so than the top). The graph in your science notebook can help you predict the weight that multiple strands will bear.

Now Try This! *Using your graph, can you predict how much weight 10 pieces of spaghetti can support? Test it out and compare your prediction to the actual number. Try out other bridge designs. Use mini marshmallows or hot glue to attach your spaghetti pieces to each other in different ways. What is the strongest bridge design you can create?*

ROCK-HOPPING ROBOT

LEVEL OF DIFFICULTY: MEDIUM

FROM BEGINNING TO END: 45 MINUTES

OTHER CATEGORIES: SCIENCE, TECHNOLOGY

THE REAL QUESTION:

Can square wheels move a vehicle? Form a hypothesis about how the size of a square wheel will affect driving motion. Use the engineering design process to create a rubber-band-powered vehicle with square wheels.

⚠ CAUTION: Be careful when using the pencil to poke holes.

MATERIALS:

- Cup or lid, about 2 inches (≈5 cm) in diameter, for circle template
- Corrugated cardboard, 1 foot (≈30 cm) by 2 feet (≈61 cm)
- Scissors
- Ruler
- 2 sharpened pencils
- Vehicle body material (cardboard tube, plastic bottle, small cardboard snack box, or another material with open space in the center and a width at least 1 inch [≈2.5 cm] smaller than the length of a pencil)
- 2 to 5 rubber bands (size #12, 14, or 16, about 2 inches [≈5 cm] by ¹⁄₁₆ inch [≈2 mm])
- Binder clip, ¾ inch (≈19 mm) to 1¼ inch (≈32 mm) wide
- Masking tape or clear packing tape

THE STEPS:

1. Use the circle template to trace and cut out two front wheels from corrugated cardboard. Cut out four wheels if you want to double layer each wheel to make it thicker.

2. Create the back wheels. Use the ruler to trace two 6-inch (≈15-cm) cardboard squares, two 4-inch (≈10-cm) cardboard squares, and two 2-inch (≈5-cm) cardboard squares. Cut out the six squares. Draw diagonal lines to connect opposite corners on the squares to form an "X." Poke a small hole where the lines cross in the center with a sharpened pencil.

3. Attach the wheel axles to the vehicle body. Use the pencil point to create holes in the front of the vehicle. Push the pencil through the holes in the front. Spin the axle (pencil) so that it moves freely in the hole. Repeat to add the back axle.

CONTINUED →

4. Attach the rubber-band power mechanism. Loop two to three rubber bands to create a chain about the length of the vehicle. Loop one end of the rubber band chain around the back axle pencil. Stretch and clip the other end of the rubber band chain to the front of the vehicle using the binder clip.

5. Attach the wheels. Tape the two 2-inch (≈5-cm) cardboard circle wheels to the ends of the front axle pencil. Tape the two 6-inch (≈15-cm) cardboard squares to the back axle pencil.

6. Spin the back axle to wind the rubber band. Let go and watch it move! Record the distance your vehicle travels in your science notebook.

7. Figure out an optimal number of times to spin the back axle. Be careful as overspinning the axle can cause too much tension and break the rubber band.

8. Remove the back wheels and repeat using the 4-inch (≈10-cm) and 2-inch (≈5-cm) square wheels. Remember to spin the back axle the same number of times for each trial.

OBSERVATIONS:

- How does the motion of the 2-inch (≈5-cm), 4-inch (≈10-cm), and 6-inch (≈15-cm) square wheels compare? How well does your vehicle travel over uneven surfaces?

The Real Science Behind How and Why: The further from the axle the edge of the wheel is, the more torque it takes to move. With the small wheels, even though the wheels are square, the energy stored in the rubber bands provides enough torque to move the wheels. As the size of the wheels increases, the amount of torque required to overcome the square corners increases until the rubber band does not have enough energy to provide spinning motion.

Now Try This! *Try using different sizes of rubber bands. Which rubber band size results in the vehicle moving the greatest distance? Redesign your vehicle so it works better. The engineer design process is prototype, test, and repeat!*

SERIES VERSUS PARALLEL STRING LIGHT CIRCUITS

LEVEL OF DIFFICULTY: MEDIUM

FROM BEGINNING TO END: 60 MINUTES

OTHER CATEGORIES: TECHNOLOGY, ART

THE REAL QUESTION:

Series circuits have just one path for electricity to flow, whereas parallel circuits have multiple paths for electricity to flow. How does this difference affect how light bulbs function in the two circuit types? Form a hypothesis about how circuit type relates to light bulb brightness. Build circuits using string lights to compare series and parallel circuits.

> **!** **CAUTION:** Have an adult help with using wire cutters or scissors to strip the wires.

MATERIALS:

- **String lights with at least 3 working light bulbs (incandescent bulbs, not LEDs)**
- **Wire strippers (or scissors)**
- **Aluminum foil, cut into three 1-inch (≈2.5-cm) by 3-inch (≈7.5-cm) strips**
- **9-volt battery**

THE STEPS:

1. Cut three lights out of the string, leaving the light bulb and about 3 inches (≈8 cm) of wire on either side.

2. Carefully strip the plastic insulation about ½ inch (≈13 mm) from the ends of each wire using wire strippers or scissors. It may take some practice to cut away just the plastic and leave the center metal wire.

3. Connect the end of the wire of one light bulb to the end of the wire of a second light bulb and tightly cover with aluminum foil. Touch each of the two loose ends of wire to one of the snap connectors on the top of the 9-volt battery. This is a series circuit. Observe the brightness of each bulb. Repeat this setup, but add a third light bulb to the row.

CONTINUED →

4. Now wire the two lights in a parallel circuit. Wrap aluminum foil around two loose wire ends, one from each side of both light bulbs. Touch this to one of the snap connectors on the top of the 9-volt battery. Wrap aluminum foil around the other two loose wire ends and touch to the other snap connector. Observe the brightness of each bulb. Repeat this setup, but add a third light bulb in parallel by touching one of its loose wire ends to one snap connector and the other loose end to the other snap connector. Tip: The two or three loose wire ends of the light bulbs can be wound together to form one combined wire to touch to a snap connector. Then, the other two or three wire ends can also be wound together to form a second wire to touch the other snap connector. This is quicker than using the aluminum foil.

OBSERVATIONS:

- ➔ How does the brightness of the bulb compare in the series and parallel circuits using two and three lights? What happens when one or more bulbs are disconnected from the circuit?

The Real Science Behind How and Why: Electrical current in a wire and light bulb can be compared to water flowing through a pipe. Series circuits are like running multiple pipes end to end. All the water must flow through the first pipe before entering the second pipe. If you block the flow in the first pipe, no water ever reaches the second pipe. This is what happened when one light was removed in the series circuit. Nothing can flow.

Parallel circuits are like pipes lined up side by side. This way, there are multiple paths for water to flow. If the flow of water stops in the first pipe, it can still flow in any other pipe. This is why removing one of the lights from the socket in the parallel circuit did not affect the other light bulbs.

Now Try This! *Apply your knowledge of circuits to build something. Try building a birthday card with the light bulbs poking through the front that turns off when the card is closed. Fold an origami firefly and place a light bulb on its tail. How can you design it so that you can press the wing to light the firefly?*

Chapter 5

Art

The "A" in STEAM technically refers to arts and humanities. The arts and humanities include physical arts (dance), fine arts (painting, drawing, sculpting), performing arts (theater, music), language arts (reading, writing), and history. The "A" in STEAM stands for applying the arts and humanities to real-world situations.

Empathy and creativity are the elements of humanity that the "A" in STEAM add. This is what makes solutions that humans create different than the ones computers create.

Art is more than just painting and drawing or creative thinking. Art is the expression of imagination and the application of creativity and empathy. Empathy involves thinking about and understanding how others feel. Empathy is an important part of a problem-solving process called design thinking. Design thinking focuses on understanding how others feel in order to find innovative solutions to problems. There are many different ways to practice design thinking,

but they all generally include the following steps: (1) empathize, (2) define the problem, (3) brainstorm ideas, and (4) test the best ideas to find a solution. Design thinking is a lot like the scientific method, but instead of a focus on figuring things out, the focus is on helping people.

In these experiments, you will use art to investigate the science behind momentum, science to investigate the art of printing photographs, and engineering to build a machine to make art. Art is applied in different ways in each experiment, but they all are examples of STEAM in action.

You'll need some art supplies for the experiments in this chapter. You might need to replenish your supply of tempera paint and watercolors. Also, add spinach to the grocery list because you will need it for the Chlorophyll Printing experiment on page 99. Get ready to integrate art and science!

MOMENTUM DROP PAINTING

LEVEL OF DIFFICULTY: EASY

FROM BEGINNING TO END: 30 MINUTES

OTHER CATEGORIES: SCIENCE, **MATH**

THE REAL QUESTION:

Momentum is the measurement of an object in motion. How does dropping a ball from different heights influence the size of paint splatter? Form a hypothesis about the patterns of splatter you think you will see. Explore the impacts of momentum while creating a work of art!

⚠ CAUTION: This experiment is messy. Wear clothes you can get paint on. This experiment is best performed outdoors.

MATERIALS:

- ◉ **Coffee mug**
- ◉ **Thin cardboard (cut open empty cereal box)**
- ◉ **Scissors**
- ◉ **Pen or marker**
- ◉ **Quarter**
- ◉ **Large piece of paper at least 2 feet (≈61 cm) by 2 feet (≈61 cm) (poster board, white side of wrapping paper, or white butcher paper)**
- ◉ **Small ball like a baseball, tennis ball, or bouncy ball**
- ◉ **Washable paint**
- ◉ **Plastic wrap (optional)**
- ◉ **Ruler, measuring tape, or meter stick**

THE STEPS:

1. Use the bottom of a coffee mug as a template to trace three (or more) circles into the cardboard. Cut out each of the cardboard circles.

2. Trace around the quarter in the center of each cardboard piece. This circle is the guide to fill with paint. This way each trial of the experiment uses the same amount of paint.

3. Lay your piece of paper outside on a flat, hard surface like a driveway.

4. Choose at least three different drop heights. Make sure each distance is still close enough that you can drop the ball carefully. The ball needs to fall directly onto the center of the cardboard circle each time.

5. Fill the 1-inch (≈1.5-cm) diameter center circle with paint. Flip the cardboard circle over and place it onto the paper, paint side down.

6. Cover the ball with plastic wrap if you are worried about it getting messy. Then drop the ball carefully so it will fall directly onto the center of the cardboard circle. Measure and record the length of the paint splatter.

7. Repeat steps 5 and 6 at least two more times using a different distance each time.

CONTINUED ➜

OBSERVATIONS:

- Create a table organizing the drop height and paint splatter measurements. Use this table to make a graph. How does the drop height of the ball affect the size of paint splatter?

The Real Science Behind How and Why: Momentum is found by multiplying mass times velocity. Momentum describes the strength of a moving object. Lightweight objects moving slowly have less momentum than heavy objects moving fast. Imagine you are at bowling alley, trying to knock down pins with a ping-pong ball. The ping-pong ball is too light no matter how fast it goes. It does not have enough momentum to knock down the pins. In this experiment, you dropped a ball from increasing heights. From the greatest height, the ball was able to reach greatest velocities each time before hitting the ground. Greater velocity made greater momentum! Jackson Pollock was a famous artist known for pouring, dripping, and splashing paint, similar to the method used in this experiment. Research "action painting" to learn more about this painting technique!

Now Try This! *Repeat the experiment using different amounts of paint and/or different balls. How does dropping different balls from the same height impact the paint splatter? How far can you get the paint to splatter if you hit the cardboard circle with a hammer?*

BUILDING AN ADVANCED DRAWING MACHINE

LEVEL OF DIFFICULTY: HARD

FROM BEGINNING TO END: 60 MINUTES

OTHER CATEGORIES: SCIENCE, TECHNOLOGY, MATH

THE REAL QUESTION:

A harmonograph is a machine that uses pendulum motion to draw a geometric picture. This experiment uses **harmonic motion**. Form a hypothesis about how pendulums can be used to create mathematical drawings. Learn about applications of pendulums by building and using a harmonograph.

! CAUTION: Have an adult help tape items to the top of the door frame.

MATERIALS:

- **40 feet (≈12 m) cotton string**
- **Scissors**
- **Measuring tape**
- **6 wooden craft sticks, regular or jumbo size**
- **Masking tape**
- **Clipboard**
- **Straw**
- **Paper clip, regular or jumbo size**
- **Plastic cup**
- **Wooden dowel, 12 inches (≈30 cm) to 18 inches (46 cm) in length**
- **Felt-tip markers**
- **Clip-type clothespin**
- **3 to 5 books**
- **Paper**

THE STEPS:

1. Choose a doorway in which to build your harmonograph.

2. Cut four pieces of cotton string, long enough to be about 6 inches (≈15 cm) taller than the height of your doorway. Make sure the length of each string is the same.

3. Tie each string to the center of a wooden craft stick and securely tape the four craft sticks to the top of your doorframe, two each on opposite sides. Space out where you tape the craft sticks such that the cotton string is similar in spacing for the four corners of the clipboard.

CONTINUED ➜

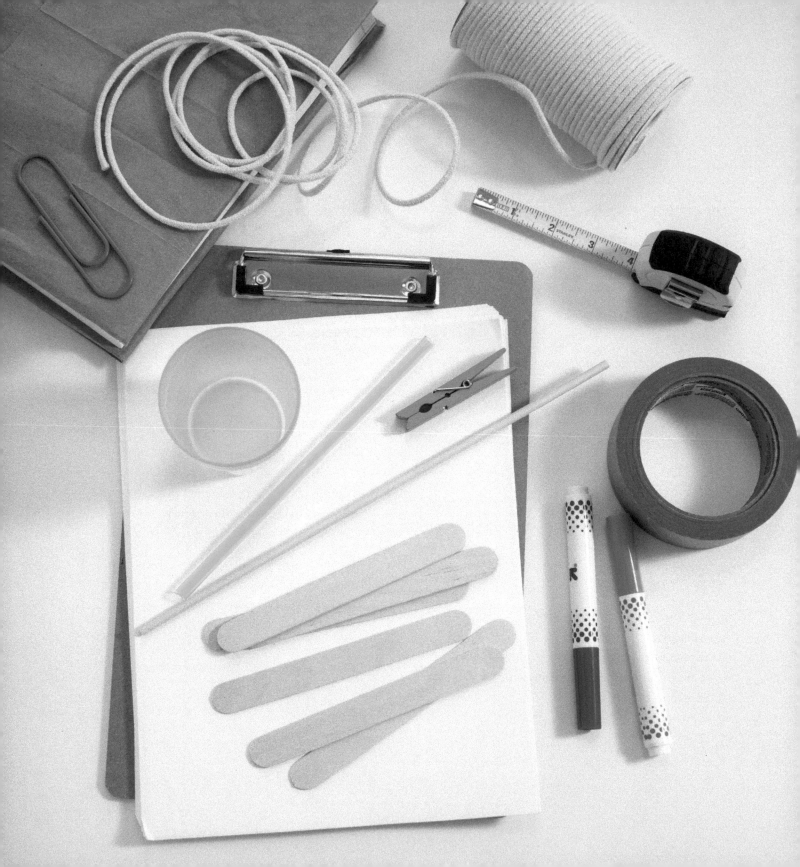

4. Securely tape the other ends of the cotton string to the corresponding four corners of the clipboard such that the clipboard hangs about 12 inches ($\approx$30 cm) above the floor, and is parallel to the floor. Tie each set of two strings into a loop underneath the clipboard before taping to add stability.

5. Cut the straw so you have one piece about 2 inches ($\approx$5 cm) in length. Straighten a paper clip and thread it through the straw hole. Center the straw on the paper clip. Bend each side of the paper clip down near each straw end to hold the straw in place in the center of the paper clip. Don't make it too tight! Make sure the straw can roll freely on the paper clip. Tape each end of the paper clip to a wooden craft stick. This is the hinge assembly.

6. Place the rim of the plastic cup down on a flat surface upside down (with the bottom facing up). Tape the two wooden craft stick legs of the hinge assembly securely to the sides of the cup at the narrowest part.

7. Tape the center of the wooden dowel to the straw on the hinge assembly at a 90-degree angle. On one side, tape a marker such that its drawing point faces the ground.

8. Tape a second marker to the end of the clothespin opposite its clip mechanism. This will serve as a counterweight to the other marker already taped to the hinge assembly. Try clipping the counterweight to different places on the wooden dowel until it balances with the other marker.

9. Place the book to add weight to the hanging clipboard. Put a piece of paper on top of the book for drawing. Tape the paper down to hold it in place.

10. The tip of the marker needs to be the same height as the paper on the clipboard. Add some books on the floor next to the clipboard's long side, with the cup hinge assembly sitting on top of the books, until the marker tip is the right height to touch the paper.

11. Move the books and the cup hinge assembly closer to the hanging clipboard so that the marker hangs near the middle of the paper. Take off the marker cap and snap it onto the other end of the marker. (This keeps it the same weight as the other marker.) You may need to adjust the position of the counterweight so that the marker rests gently on the paper.

12. Gently swing the clipboard, and let the drawing begin!

CONTINUED ➜

OBSERVATIONS:

⊙ **How do the drawings compare when you push versus gently twist the hanging clipboard?**

The Real Science Behind How and Why:
The motion of the plane (where the paper is being held) creates a pattern based on the pendulum motions of each string holding the clipboard. If these moved in one direction (such as only back and forth), then it would draw a straight line. Since the clipboard is moving in two directions (back/forth *and* left/right), complicated pictures are drawn based on small changes in the pendulum motion. The way a pendulum moves is a special type of motion called harmonic motion. The harmonic motion of a pendulum is a repetitive back and forth movement that continues due to the force of gravity. Eventually the motion of the harmonograph stops due to **friction** from the marker moving on the surface of the paper.

Now Try This! *Using a paper clip as a hook, hang bags of pennies of each corner of the hanging clipboard and then swing it again. Try using the same number of pennies in each bag or a different number of pennies in each bag. How do the pennies change the drawing motion of the machine?*

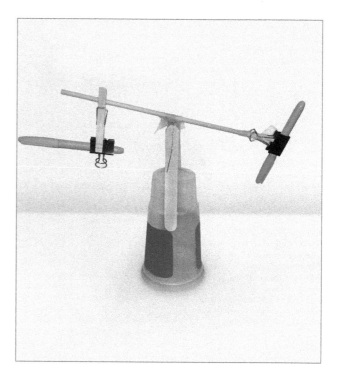

CHLOROPHYLL PRINTING

LEVEL OF DIFFICULTY: EASY

FROM BEGINNING TO END: 2 TO 5 HOURS

OTHER CATEGORIES: SCIENCE, TECHNOLOGY

THE REAL QUESTION:

How can you make prints using natural pigments such as **chlorophyll** and the sun? Form a hypothesis about how the number of coats of pigment affects the quality of the print. Learn about the power of the sun by processing your own prints.

MATERIALS:

- 2 or more cups spinach
- Blender
- Spoon
- Fabric for strainer (old T-shirt, old sock, cheesecloth)
- Large bowl
- Foam brush
- Paper (thick watercolor paper works best, but regular paper works, too)
- Thin items such as flattened leaves or flowers
- Clear plastic sheets (from report covers or picture frames)
- Binder clips

THE STEPS:

1. Put the spinach into the blender and grind into a fine pulp.

2. Scoop the pulp into a section of the strainer fabric. Squeeze the fabric and pulp tightly over a bowl until all the liquid comes out. Add the pulp a couple scoops at a time to the strainer fabric and use a wringing motion. This will be your pigment.

3. Use the foam brush to paint a layer of the pigment onto the paper. Let it dry for about 5 minutes and add another coat. Create one sheet that has only one coat, another that has two coats, and a third that has three coats of pigment.

4. Lay one or two thin, flat items on each of the three pieces of paper.

5. Sandwich each paper between two plastic sheets and clamp with binder clips.

6. Put out in direct sunlight around noon for 1 to 4 hours. Check the print every hour by carefully moving the item to see how much the color has changed around the item compared to underneath the item. Remove the print from the sun once the desired amount of color change has occurred.

CONTINUED →

OBSERVATIONS:

⊙ **How does the print using one coating of pigment compare to two or three coats?**

The Real Science Behind How and Why: This type of printing recreates one of the first type of photographs, called anthotypes. This method creates a silhouette using photosensitive material from plants exposed to sunlight. Different plant pigments need different exposure times, ranging from hours to weeks. Chlorophyll is a green pigment found in plants that captures energy from sunlight, helping in the process of photosynthesis. Photosynthesis is how plants convert light energy into chemical energy.

Now Try This! *Try using other plants to create photosensitive pigment, such as colorful flower petals, cabbage, or kale. What happens if you try to make prints at different times of day?*

GEOLOGICAL ART

LEVEL OF DIFFICULTY: EASY

FROM BEGINNING TO END: 2 DAYS
(ALLOW TO DRY OVERNIGHT TWICE)

OTHER CATEGORIES: SCIENCE

THE REAL QUESTION:

How do water and color absorb into different **geological** materials? Form a hypothesis about how watercolor will flow and absorb over and into your natural art piece. Create a mosaic artwork inspired by natural materials.

! CAUTION: Make sure all material is free of living creatures.

MATERIALS:

- Rocks
- All-purpose white glue
- Base for artwork, such as scrap wood or cardboard
- Sand
- Natural materials such as soil, salt, twigs, feathers, flowers, or leaves
- Paintbrush or water dropper
- Watercolors or watered-down food coloring

THE STEPS:

1. Think about a concept for your mosaic artwork. It can be abstract art that conveys a feeling or representational art that looks like something in particular.

2. Determine placement of rocks and glue them to the base.

3. Add glue around the rocks and sprinkle sand and/or soil onto the glue.

4. Glue down any other natural materials you want to add. Allow to dry overnight.

5. Using the paintbrush or water dropper, apply watercolor over the sand. Add a small amount of additional water if needed to help the color flow. Note: Adding too much water will weaken the glue. Allow to dry overnight.

OBSERVATIONS:

- How does the watercolor flow and absorb into the different materials?

The Real Science Behind How and Why:
The surface of Earth is dynamic, meaning it is always changing. Weathering breaks apart the surface of Earth. Physical weathering is a result of a physical change, like water freezing in the crack of a rock. Chemical weathering changes the chemical composition of a rock. For example, when acidic rainwater flows over granite, it causes a chemical change that turns the mineral feldspar into clay. Erosion due to wind, liquid water, and ice is constantly moving pieces of weathered earth material, called sediment, to new places. Deposition occurs when sediment is added to a new location. Artwork outside (such as sculptures) becomes a part of this changing process.

Now Try This! *How do the results change over time if you leave the artwork outside in full sun versus a shaded area? Take photos and add notes in your science notebook to record observations of the changes over time. Create another artwork using different combinations of salt, sugar, glue, and watercolor.*

DESIGN A HOLOGRAM

LEVEL OF DIFFICULTY: MEDIUM

FROM BEGINNING TO END: 45 MINUTES

OTHER CATEGORIES: SCIENCE, TECHNOLOGY, ENGINEERING, MATH

THE REAL QUESTION:

How can you make a shape appear to be floating in space by scratching plastic? Form a hypothesis about how the size and shape of a **hologram** will compare to its original image. Create an image using a process called scratch holography.

> ⚠ **CAUTION:** Be careful using thumbtacks. Do not shine a flashlight directly into your eyes.

MATERIALS:

- Paper
- Clipboard
- Masking tape
- CD
- 2 thumbtacks
- Wooden craft stick
- Pencil
- Flashlight

THE STEPS:

1. Put the paper in the clipboard and use two pieces of tape to attach the shiny side of the CD to the top of the paper.

2. Build an image transfer device (a compass) by carefully sticking a thumbtack all the way through each end of the wooden craft stick. Spin the point of the thumbtack slowly into the wood of the craft stick so that the wood does not split. Add tape around the end of the craft stick so that the thumbtack does not shift position.

3. Place one point of this compass just below the hole in the center of the CD. Where the other end of the compass touches the paper vertically, mark a dot, and then draw a horizontal line on the paper. Repeat this process about ⅜ inch (≈1 cm) above the bottom edge of the CD. This will make the distance between your drawn image and the CD markings align with the length of your compass.

4. Draw the design you want to make into a hologram between the lines. Make the design narrower than the CD. Start with a simple shape like a heart or block letters. Keep in mind that the hologram will end up slightly smaller than what you draw.

CONTINUED ➔

5. Using the pencil, put dots on each line of your picture spaced about every ⅛ inch (≈3 mm).

6. For each dot you just drew, place one end of the compass on the dot to act as a pivot point, then move the other end of the compass in an arc from edge to edge on the CD to lightly scratch its surface. Be careful not to dig too deep. Trace each arc one time.

7. Shine a flashlight at a high angle horizontally across the surface of the CD while tilting the plastic back and forth to see your hologram appear as the glints of shiny dots float in space above your CD. It may be tricky to see at first depending on the angle of the light, so experiment to find which angle works the best.

OBSERVATIONS:

◉ **How far left or right from the middle of the plastic can your eyes be before the hologram disappears? How does the appearance of the shape and size of the hologram compare to the original image?**

The Real Science Behind How and Why:
The angle of the light, the plastic, and your eyes create the holographic illusion. The scratch reflects light strongly at a single point along the arc based on the angle of the light source and your eye. The distance between your eyes makes each eye see things from a slightly different angle. The difference in the angle with which each eye views the scratch means that each eye sees the reflected spot of light at a different place along the arc. This gives the illusion of depth, because when looking at regular objects in the world, our brains are accustomed to using the two perspectives of each eye to calculate which objects are farther away than others. This is known as depth perception. If you close one eye, your depth perception is diminished, and your brain will have to rely on other cues (such as relative size) to tell how close objects are to you and to each other. You can play with this by closing one eye and softly tossing a tennis ball with a friend. To see high-definition examples of scratch holograms, research the specular holography work of artist Matthew Brand.

Now Try This! *Can you shine two bright flashlights from different angles and see more than one hologram at the same time? Create a new hologram with a more complicated shape, like a cube.*

Chapter 6

Math

Regardless of the first language you speak, everyone speaks math. Math is called the universal language because everyone everywhere uses it. The symbols and the ways to form equations are the same worldwide.

One way to build your understanding of math language is to create benchmarks for measurements. A benchmark is a point of reference to make comparisons. Look at the benchmarks below and find other items to represent different lengths. For example, your hand length may be about 6 inches (≈15 cm). You are carrying around a ruler and didn't even know it!

0.1 cm ≈ 1 mm ≈ width of a dime

1 cm ≈ width of your fingertip

2.5 cm ≈ 1 inch ≈ width of a quarter

30 cm ≈ 1 foot ≈ length of a piece of computer paper

100 cm ≈ 1 meter ≈ 1 yard ≈ width of door

Math is important in all fields of study. Every single experiment in this book involves math. Whether it is through patterns, volume, shapes, measurement, or some other way, math is a part of science, technology, engineering, and art. Go back and look at experiments you've already completed and see how many math connections you can make.

This chapter investigates different aspects of math, from geometry to the measurement of time. Two experiments in this chapter investigate special patterns with the golden ratio. You will even learn a special way of counting with the binary number system. Math is the language of science. You'll communicate with math in the experiments in this chapter!

USING THE GOLDEN RATIO TO MAKE A MECHANICAL TENTACLE

LEVEL OF DIFFICULTY: EASY

FROM BEGINNING TO END: 60 MINUTES

OTHER CATEGORIES: SCIENCE, ENGINEERING

THE REAL QUESTION:

The **golden ratio** is a special number with the value 1.618. How do your hand measurements compare to the golden ratio? How can you use these measurements to make a mechanical tentacle? Learn about connections in math and nature with this engineering challenge!

MATERIALS:

- Ruler
- Paper
- Straw
- Marker
- Cotton string or yarn
- Paper clip
- Calculator

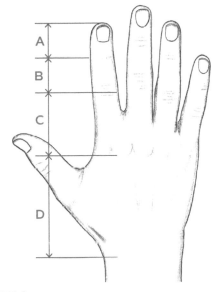

THE STEPS:

1. Measure the lengths of A, B, C, and D using your own hand. A is the length from your fingertip to the first knuckle. B is the length from the first to second knuckle. C is the length from the second to third knuckle. D is the length from the third knuckle to the point where your wrist meets your hand. Record each measurement.

2. From one end of the straw, mark these four lengths using your marker.

CONTINUED →

3. Cut a small notch in the straw at each of the four lengths.

4. Thread a string all the way through the straw. Tie a large knot in the end closest to the shortest measurement. To keep the knot from falling inside your straw, tie a paper clip to the knot end.

5. You've created a mechanical tentacle. Hold the other end of the straw closest to the longest measurement, and pull the string. Release the tension in the string.

OBSERVATIONS:

⊙ What happens when you pull and release the string? Copy and complete the chart to find the ratio of your hand measurements. How do your hand measurements compare to the golden ratio?

Copy and fill out this table on your paper:

LENGTHS	CALCULATIONS
A=	
B=	
C=	
D=	
B ÷ A=	__ ÷ __ = __
C ÷ B=	__ ÷ __ = __
D ÷ C=	__ ÷ __ = __

The Real Science Behind How and Why: The mechanical tentacle is inspired by organisms, especially the octopus. Engineers use robotic tentacles to pick up objects that are tricky to hold due to their complex or easy-to-break shape. Engineers have designed the spacing of mechanical joints to follow the golden ratio. Your hand is just one example of the golden ratio. The golden ratio is also related to the pattern of seeds within a sunflower and the arrangement of leaves on a plant stem. Research the golden ratio to find more cool examples of this number!

Now Try This! *Complete this experiment using the hand measurements of friends and family members. Whose measurements are closest to the golden ratio? Whose tentacle works the best? For a fun challenge, make eight tentacles, tie them to a paper cup body with string, and create an octopus!*

CREATING TIME

LEVEL OF DIFFICULTY: EASY

FROM BEGINNING TO END: 15 MINUTES

OTHER CATEGORIES: SCIENCE, TECHNOLOGY

THE REAL QUESTION:

How can a pendulum be used to create a clock? The time for one complete swing (a right swing and a left swing) is called the period. Form a hypothesis about the length of string needed for a complete swing to have a period of exactly one second. Learn about gravity with this pendulum investigation.

MATERIALS:

- Masking or packing tape
- Pencil
- Ball of clay, about 1 inch (≈2.5 cm) in diameter
- 2 feet (≈0.7 m) to 3 feet (≈0.9 m) of cotton string or yarn
- Binder clip
- Ruler
- Stopwatch

THE STEPS:

1. Tape a pencil to a table or another tall surface such that it is hanging halfway over the edge.

2. Roll a ball of clay into one end of the string.

3. Clamp the other end of string to the overhanging pencil using the binder clip. Use the clamp to adjust the length of the string to the desired height for each trial. Try 6 inches (≈15 cm), 12 inches (≈30 cm), and 18 inches (≈46 cm) for your first three trials.

4. Pull the clay ball and let go. Measure the time for 10 complete swings. Divide this time by 10 to find the period. For example, if it takes 12 seconds for 10 complete swings, one complete swing is 1.2 seconds.

OBSERVATIONS:

- What length of string creates a complete swing that is exactly one second?
- Does a longer pendulum have a longer or shorter period than a shorter pendulum? If the pendulum length is twice as long, is the time for a complete swing twice as long?

CONTINUED →

The Real Science Behind How and Why:
Everything on Earth is pulled towards the ground at the same rate of 9.8m/sec^2, which is acceleration due to gravity. The pendulum is pulled downward at this same rate no matter what size the clay ball is. A pendulum with a period of 1 second will have a length of about 25 cm. Length is the most important factor for a pendulum's period because a longer pendulum has farther to fall, creating the longer back-and-forth swing of the pendulum.

Now Try This! *Keep the length of the string the same, but change the size (and therefore mass) of the clay ball. How does changing the mass of the clay ball impact the period?*

WEAVING A MAGIC SQUARE

LEVEL OF DIFFICULTY: EASY

FROM BEGINNING TO END: 30 MINUTES

OTHER CATEGORIES: ART, TECHNOLOGY

THE REAL QUESTION:

A magic square is a square grid filled with consecutive numbers arranged in such a way that the sum of the numbers in each row, column, and diagonal is the same. This sum is called the magic constant. How can you create a magic square? Form a hypothesis about how the numbers 1 to 16 can be arranged to create a magic square.

MATERIALS:

- Paper
- Scissors
- Pencil or pen

1	2	3	4
5	6	7	8
9	10	11	12
13	14	15	16

+

16	15	14	13
12	11	10	9
8	7	6	5
4	3	2	1

=

1	15	14	4
12	6	7	9
8	10	11	5
13	3	2	16

THE STEPS:

1. Cut the paper into two 8½-inch (≈22-cm) squares. Fold each square in half and in half again in both the horizontal and vertical directions. The folds in the paper should form a 4 × 4 grid.

2. In the first 4 × 4 grid, write the consecutive numbers 1 to 16 starting in the top left corner and continuing horizontally to fill each row before moving down one to the next row.

3. In the second 4 × 4 grid, write the consecutive numbers 1 to 16 starting in the bottom right corner and continuing horizontally to fill each row before moving up one to the next row.

4. For both 4 × 4 grids, you will make two cuts to create three strips. You will only cut three boxes deep. Cut the first strip so it is one box wide, the second strip so it is two boxes wide, and the last strip so it is one box wide.

5. Weave the three strips into the pattern shown in the diagram. You have created a magic square!

OBSERVATIONS:

- **What is the magic constant of this magic square? How many other patterns in the 4 x 4 square can you find? Aside from adding the diagonals, rows, and columns, how can you find the magic sum?**

The Real Science Behind How and Why:
The magic square you created is the same one made famous by the German artist and mathematician Albrecht Dürer in 1514. The rows, lines, diagonals, and four corners in this square all add up to 34. Even the four digits in the square at the center, top right, top left, bottom right, and bottom left all add to 34! The Dürer's magic square is linked to a special way to send a secret message called the pigpen cipher. The pigpen cipher uses fragments of a grid to exchange letters for symbols. In earlier times, cipher codes were used to write letters and keep records so others could not easily read them.

Now Try This! *What patterns can you find if you weave the strips in a different way? How can you use each of the numbers 1 to 9 once to create a 3 x 3 magic square? (Hint: The magic sum will be 15 for a 3 x 3 square.)*

ICOSAHEDRAL VIRUS

LEVEL OF DIFFICULTY: HARD

FROM BEGINNING TO END: 30 MINUTES

OTHER CATEGORIES: SCIENCE, ENGINEERING

THE REAL QUESTION:

A polyhedron is a solid geometric figure with many flat faces, like a pyramid or a cube. A 20-sided polyhedron is called an icosahedron. How can you use 20 equilateral triangles to create a 20-sided shape? Form a hypothesis about how to organize 20 equilateral triangles to build an icosahedron. Learn about the common icosahedral structure of viruses by building your own viral packaging.

MATERIALS:

- **Geometric drawing compass (same from the one from Design a Hologram experiment on page 105)**
- **Thick paper (8½-inch [≈22-cm] x 11-inch [≈28-cm] card stock or stiff paper)**
- **Ruler**
- **Pencil**
- **Scissors**
- **Transparent tape**
- **Material for filling (enough tissue paper, toilet paper, or cotton balls to fill your shape, about the size of a small fist)**

THE STEPS:

1. Create a template for making 20 equilateral triangles. An equilateral triangle is a triangle in which all three sides are equal and all three inside angles are 60 degrees. Use a ruler to draw a line segment that is about 1½ inches (≈38 mm) in length. Adjust your compass so it is the same width as the line segment you drew, with one side of the compass touching the first point (point A) and the other side of the compass touching the second point (point B).

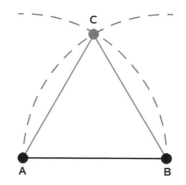

2. Without changing the size of your compass, put your compass on point A and draw an arc. Repeat on point B. Color a dot where the two arcs meet (point C). Use a ruler to draw a straight line from point A to point C and point B to point C. Carefully cut out this triangle to use as a template. (If you prefer, you can use the equilateral triangle template).

3. Trace your triangle template onto the thick paper 20 times and cut out 20 identical triangles.

4. Figure out a way to tape the triangles together so each of their edges completely touch and enclose the soft material.

5. The soft material helps the icosahedron you are building to maintain its shape, so add it to the inside of your shape in small amounts as needed while you are experimenting with different ways to build your shape.

OBSERVATIONS:

- Record a description of your final shape in your science notebook. How many sides, edges, and corners are on your icosahedron?

The Real Science Behind How and Why:
Viruses are microscopic parasites that infect plants, animals, fungi, and bacteria. Viruses are much smaller than bacteria and can only replicate inside the living cells of other organisms. Viruses come in a variety of shapes and sizes. The outermost layer of a virus is a protein shell called a capsid. Most viruses have capsids that are helical or icosahedral in shape, but they can come in other complex shapes, too. An icosahedral shape has 20 triangular faces, 30 edges, and 12 corners called vertices. This shape helps a virus decrease its ratio of surface area to volume so it can carry more genetic material within its shell.

Now Try This! *Use your model to create a die with 20 sides. Fair dice have an equally likely chance of landing on any face. Design a probability experiment to determine if your 20-sided dice would be considered fair.*

COMPLEX COUNTING CHALLENGE

LEVEL OF DIFFICULTY: EASY TO HARD

FROM BEGINNING TO END: 30 MINUTES

OTHER CATEGORIES: TECHNOLOGY

THE REAL QUESTION:

The base of any number system is how many unique symbols are used to represent numbers. The decimal number system is base 10 because it uses ten symbols to represent the numerals 0 to 9. The binary number system is base 2 because it uses only two symbols (0 and 1). How can you represent the binary number system using just your hands? Form a hypothesis about how high you can count using binary on one hand. Think like a computer and learn how to count way past ten on just one hand. Binary is as easy as 1, 10, 11! (1, 10, and 11 are binary for the values 1, 2, and 3.)

MATERIALS:

- Paper
- Pencil

THE STEPS:

1. Place your right hand with your palm facing up. Assign each finger position the values of 1, 2, 4, 8, and 16 from right to left. This is similar to the 10s place, 100s place, and so on in the decimal number system, but instead you have the 1s place, 2s place, 4s place, 8s place, and 16s place.

2. A finger in the down position represents "0" and a finger in the up position represents "1." For example, the value of 1 is easily shown by just having your thumb up to represent the 1's place. Similarly, just put up your pointer finger for the value of 2, your middle finger for the value of 4, your index finger for the value of 8, or your pinky for the value of 16.

3. The values of other numbers are represented in the binary system by adding the value of each position as needed. For example, the value of 3 is shown by having your thumb up to represent the 1's place plus your pointer finger to represent the 2s place (1 + 2 = 3).

4. Copy and complete the chart below. Continue counting as high as you can using one hand. Add and fill out another row in the chart for each additional decimal number. When all your fingers are in the up position you have represented the greatest value you can on one hand.

OBSERVATIONS:

- What patterns do you notice in the sequence of numbers? What comparisons can you make between the base 10 and base 2 number systems? How high can you count on just one hand?

Now Try This!

Continuing from right to left, assign the positions on your left hand, palm up. The next set of values continue the binary pattern: 32, 64, 128, 256, and 512. Now, how high can you count on both hands?

HAND							
	Pinky Finger	Ring Finger	Middle Finger	Pointer Finger	Thumb		
Decimal Number	16	8	4	2	1	Made By	Binary Number
1	down	down	down	down	up	1	1
2	down	down	down	up	down	2	10
3	down	down	down	up	up	2+1	11
4	down	down	up	down	down	4	100
5	down	down	up	down	up	4+1	101
6	down	down	up	up	down	4+2	110
7	down	down	up	up	up	4+2+1	111
8						8	1000
9	down	up	down	down	up		1001
10	down	up	down	up	down	8+2	1010
11							
12							1000

The Real Science Behind How and Why: In binary, each position has the value of 2 to a growing exponential power ($2^0 = 1$; $2^1 = 2$; $2^2 = 4$; $2^3 = 8$; $2^4 = 12$), so each position is double the prior one. You can count from 0 to 31 on one hand using binary: 00001, 00010, 00011, 00100, 00101, 00110, 00111, 01000, 01001, 01010, 01011, 01100, 01101, 01110, 01111, 10000, 10001, 10010, 10011, 10100, 10101, 10110, 10111, 11000, 11001, 11010, 11011, 11100, 11101, 11110, and 11111.

Computers work a lot like this counting method. A microchip has multiple electrical switches that are either on (represented by "1") or off (represented by "0"). By just using only 1s and 0s, computer microchips can perform potentially infinite operations! Other common number systems include base 8 (octal) and base 16 (hexadecimal). Research number systems to learn more, including why some people think we should use base 12 (duodecimal) instead of a base 10 (decimal) number system. Which historical civilizations used number systems that resulted in our having 60 minutes in an hour, 24 hours in a day, and 360 degrees to draw a circle?

GINORMOUS TETRAHEDRAL CHALLENGE

LEVEL OF DIFFICULTY: EASY

FROM BEGINNING TO END: 45 MINUTES

OTHER CATEGORIES: ENGINEERING

THE REAL QUESTION:

A tetrahedron is a triangular pyramid. How can you build a structure completely out of tetrahedrons? Form a hypothesis about the size and design of your structure using 120 twelve-inch pieces. Apply math skills with this engineering design challenge.

⚠ CAUTION: Be careful with the point on the bamboo skewer. Rubber bands may break when stretched.

MATERIALS:

- 120 bamboo skewers, 12 inches (≈30 cm) long
- 200 rubber bands, 2 inches (≈5 cm) long by 1/16 inch (≈2 mm) wide (size #12, 14, or 16)
- Ruler

THE STEPS:

1. Use three skewers to form a flat triangle shape. Wrap a rubber band around the two skewers that form each corner of the triangle until tight. Using this triangle as the base, attach another skewer to each corner with a rubber band. Gather the other end of these three new skewers and bring them to a point to form the top of the triangular pyramid. Secure with a rubber band. This forms one tetrahedron.

2. Repeat this process to build more tetrahedrons until you have enough tetrahedrons for your design.

3. Attach the tetrahedrons together using rubber bands. Each of the three corners of the base of the tetrahedron can be attached to the top point of another tetrahedron. Repeat this process until you use all your individual tetrahedrons.

OBSERVATIONS:

- What are the dimensions of your completed structure? What patterns can you find in the number of triangles and skewers in your structure?

CONTINUED ➔

The Real Science Behind How and Why:
One application of the tetrahedral shape is in aeronautics. Alexander Graham Bell is best known for first patenting the telephone, but he also made advances in flight. Using a tetrahedral shape, he was able to build increasingly larger kites without increasing the ratio of weight-to-surface area. Research tetrahedral kites to learn how to build your own.

Now Try This! *What other ways can you use these same materials to build a structure? What is the tallest structure you can build?*

SHOOTING STRAWS

LEVEL OF DIFFICULTY: MEDIUM

FROM BEGINNING TO END: 45 MINUTES

OTHER CATEGORIES: SCIENCE, ENGINEERING

THE REAL QUESTION:

How does launch angle affect projectile motion? Form a hypothesis about how the launch angle affects the distance a projectile travels. Launch straw rockets to learn about projectile motion.

> **CAUTION:** Rubber bands can break unexpectedly. Do not aim rubber bands or rockets at other people, animals, or breakable objects.

MATERIALS:

- Rubber bands, ¼ inch (≈6 mm) wide (#62, 63, or 64)
- Transparent tape
- 7- to 8½-inch (≈18- to 22-cm) straight plastic straw
- Clay
- Wooden skewer
- Protractor
- Ruler
- Calculator

THE STEPS:

1. Place a rubber band over the end of the straw. Wrap tape around the straw and rubber band to secure the rubber band in place.

2. You will launch the straw rockets from the floor. Form the clay into a 2- to 3-inch (≈5- to 8-cm) diameter ball. Stick the pointy end of the wooden skewer into the ball of clay at a 30-degree angle to the floor.

3. Place the straw over the flat, blunt end of the skewer. Use one hand to hold where the bamboo stick meets the clay and the other hand to pull back the straw. Pull the straw back about one or two inches and let go. Shoot a few test launches to figure out an optimal distance to pull the straw. Record this measurement and use this pull-back distance for all the launches for this experiment.

CONTINUED →

4. Measure and record the distance the straw rocket travels from launch point to impact point for three trials. Make sure to measure to where the straw first hits the floor, not how far it slides after making impact. It is easier if a partner watches the projectile carefully and then quickly puts a finger on the ground where it impacts so you can measure to the exact spot.

5. Repeat steps 2 to 4 using a 45-degree launch angle and a 60-degree launch angle.

OBSERVATIONS:

⊘ **Calculate the average distance traveled for each launch angle by adding the three distances and dividing by three. Which angle resulted in the greatest distance traveled?**

The Real Science Behind How and Why: A projectile is any object propelled through the air. The distance a projectile travels is its range. Projectiles must move against the force of gravity in the vertical direction and the force of air friction (drag) in the horizontal direction. A projectile launched directly vertical moves at a 90-degree angle, and one launched straight horizontally moves at a 0-degree angle. As you discovered, a projectile launched exactly between these two angles (45-degree angle) perfectly balances the horizontal and vertical forces.

Now Try This! *Use tape to add fins to the straw rocket. Which fin shape, size, and arrangement results in the greatest distance traveled when your rocket is launched?*

Chapter 7

Putting It All Together

Wow, you have made it to the end of this book! However, it really is not the end. It is just one step in your scientific journey of learning and discovery. Along the way you have completed experiments that integrate Science, Technology, Engineering, Art, and Math (STEAM) and you should now have a greater understanding and appreciation for each of these subjects. You have seen the **synergy** that these concepts produce when used together. For example, the images created by the Building an Advanced Drawing Machine experiment on page 95 (technology) are beautiful and complex (art) and are created by the calculations (math) of pendulum motion (science). As smart as you are, you are already noticing these different aspects in your daily life and in school more clearly.

You have developed skills used by real scientists—the process of the scientific method—to make your own discoveries. What were the most interesting discoveries you made while using this book? What were some of your most challenging moments and why? What will be the next step in your STEAM learning journey?

STEAM requires Critical Thinking, Communication, Creativity, and Collaboration, which together are known as "the four Cs." These four skills are important in every single career. Wherever your passion and motivation take you, STEAM skills will help you to succeed. Stay curious and keep writing in your science notebook. Continue asking questions and figuring out how to make things better. You have made it this far. Keep going!

GLOSSARY

acoustical engineering: the application of sound used in technology

anaglyph: a picture with two different-colored images (usually red and blue) that appears three-dimensional when viewed using one red and one blue lens

anemometer: a common weather station instrument for measuring wind speed

angular momentum: the speed of an object moving in a circular path

barycenter: the center of mass between two or more celestial bodies in space (like the Earth and sun); the point around which celestial bodies orbit

bioglue: an adhesive that uses biological materials such as sugars or starch

biomimicry: applying inspiration from nature to create inventions that solve human problems

butterfly effect: a property of chaos theory in which small changes in initial conditions lead to large, unpredictable outcomes

centripetal force: a force directed towards the center of the circular path in which an object is moving

chaos theory: the study of systems that are very sensitive, in which very small changes in the beginning can lead to completely different outcomes

chemoreceptor: a special cell that sends a message inside an animal when it senses a certain chemical

chlorophyll: a green pigment found in plants that captures energy from sunlight, helping in the process of photosynthesis

convection current: the fluid motion of liquid or gas created as warm material rises and cool material sinks

design thinking: a creative problem-solving process that focuses on understanding how others feel in order to find innovative solutions

dynamics: a branch of physical science that studies the motion of objects and the effects from physical factors such as force, mass, momentum, and energy

electrostatic force: the attraction or repulsion of objects due to their electric charges; opposite-charged objects attract and like-charged objects repel

entomologist: a scientist who studies insects

Faraday cage: a metal box that shields electromagnetic radiation, blocking reception of a cell phone or radio

fluorescence: the glow caused when a substance absorbs ultraviolet energy and re-emits it as visible light

friction: a force that opposes the motion of an object

geological: referring to Earth in physical structure and material

golden ratio: a special ratio found in some shapes, such as the dimensions of a regular pentagon; it is used to create pleasing proportions in artwork and architecture

harmonic motion: a continuous motion due to a restoring force, such as the force of gravity causing a pendulum to continue to swing

heart rate: the number of times the heart beats in one minute

hertz: a unit which measures how often something happens; for sound, a frequency of 1 hertz means that a sound wave vibrates once a second

hologram: a three-dimensional image created with light and mirrors

hypothesis: an idea or explanation that is tested through experimentation

inertia: resistance to changes in velocity (speed)

interdisciplinary: combining one or more subjects

kinetic energy: the energy of an object that's in motion

moment of inertia: a rotating object's resistance to changing its motion

parallax: the difference in the observed position of an object when viewed from two different lines of sight

pheromone: a chemical substance made and put into the environment by an animal that influences the behavior of other members of that species

potential energy: the energy stored in an object due to its position; for example, a ball's height above ground or the distance a rubber band is stretched

protists: single-celled organisms

spatial navigation: the process of using multiple environmental cues, like landmarks or scent trails, to figure out and travel the right path to a destination

synergy: a combined effect which is greater than the sum of individual effects

thermoplastic: a type of plastic that softens when heated and hardens when cooled

variable: an aspect of an experiment that can change

FOR FURTHER INVESTIGATION

Mrs. Harris Teaches Science offers additional information about the STEAM experiments in this book plus more STEAM experiments created by the author. MrsHarrisTeaches.com

Technovation is an awesome engineering online community lets kids build, share, and get feedback on their creations. CuriosityMachine.org

Exploratorium "Snacks" are hands-on science activities that use inexpensive, readily available materials. Exploratorium.edu/snacks

Instructables has 100 STEAM projects for teachers and families to explore. Instructables.com /id/100-STEAM-Projects-for-Educators/

National Aeronautics and Space Administration (NASA) offers related articles, activities, and resources for students in grades K-12. Nasa.gov/stem /forstudents

PHET has free, interactive simulations for science and math. Phet.colorado.edu

Science Bob contains instructions and videos for interactive science experiments. ScienceBob.com

Science Journal is a free app for Android or iPhone that can measure light, sound, acceleration, air pressure, and more using a phone's built-in sensors. ScienceJournal.withgoogle.com

Scientific American offers simple and fun science experiments for families to do together in 30 minutes or less. ScientificAmerican.com/education /bring-science-home

SciStarter can help you find citizen science projects that match your interests in order to participate and collect data for scientists. SciStarter.org

REFERENCES

"14 Grand Challenges for Engineering in the 21st Century." National Academy of Engineering. Accessed September 3, 2019. http://www .engineeringchallenges.org/challenges.aspx.

"Chladni Plates." Smithsonian National Museum of American History. Accessed September 3, 2019. https://americanhistory.si.edu/science/chladni.htm.

Fellman, Megan. "Synthetic Adhesive Mimics Sticking Powers of Gecko and Mussel." Northwestern University, July 18, 2007. https://www.northwestern .edu/newscenter/stories/2007/07/messersmith.html.

Roguin, Ariel. "Rene Theophile Hyacinthe Laënnec (1781–1826): The Man Behind the Stethoscope." *Clinical Medicine and Research*, September 2006. https://www.ncbi.nlm.nih.gov/pmc/articles /PMC1570491/.

"What Is Biomimicry?" Library of Congress, July 31, 2017. https://www.loc.gov/rr/scitech/mysteries /biomimicry.html.

EXPERIMENT INDEX

Art
    Bestie Marble Maze Test, 17–19
    Building 3D Lenses, 21–23
    Building an Advanced Drawing Machine, 95–98
    Chlorophyll Printing, 99–101
    Design a Hologram, 105–106
    Dissecting Sound, 35–37
    Floating Dollar Bills, 24–25
    Geological Art, 102–103
    Homopolar Motor Color Whirl, 61–63
    Momentum Drop Painting, 93–94
    Rubber Band Power 2-in-1, 33–34
    Series Versus Parallel String Light Circuits, 87–88
    Shrinking Plastic, 27–28
    Weaving a Magic Square, 116–117

Engineering
    Bestie Marble Maze Test, 17–19
    Building 3D Lenses, 21–23
    Building with Bioglue, 67–68
    Candle-Powered Boat, 77–79
    Cell Phone Signal Blocking, 59–60
    Design a Hologram, 105–106
    Ginormous Tetrahedral Challenge, 123–124
    Homopolar Motor Color Whirl, 61–63
    Icosahedral Virus, 118–119
    Insect Mind Control, 7–8
    Light Versus Sound, 48–49
    Make a Microscope, 43–44
    Make Your Own Molds, 69–71
    The Original Fidget Spinner, 38–39
    Put a Cork in It, 75–76
    Rock-Hopping Robot, 83–85

Rotational Races, 29–31
Rubber Band Power 2-in-1, 33–34
Secret Patterns of Sound Waves, 45–47
Seeing the World Through a Bubble, 12–13
Series Versus Parallel String Light Circuits, 87–88
Shooting Straws, 125–127
Spaghetti Strength, 81–82
Using the Golden Ratio to Make a Mechanical
    Tentacle, 111–112
Walk-Along Spinning Fish, 72–73

Math
    Building an Advanced Drawing Machine, 95–98
    Chaos and the Double Pendulum, 15–16
    Complex Counting Challenge, 120–121
    Creating Time, 113–115
    Design a Hologram, 105–106
    Floating Dollar Bills, 24–25
    Ginormous Tetrahedral Challenge, 123–124
    Hydroponic System, 51–52
    Icosahedral Virus, 118–119
    Make Your Own Molds, 69–71
    Momentum Drop Painting, 93–94
    Rotational Races, 29–31
    Shooting Straws, 125–127
    Shrinking Plastic, 27–28
    Spaghetti Strength, 81–82
    Static UFO, 9–11
    Super Simple Stethoscopes, 56–57
    Using the Golden Ratio to Make a Mechanical
        Tentacle, 111–112
    Weaving a Magic Square, 116–117

Science

    Bestie Marble Maze Test, 17–19
    Building 3D Lenses, 21–23
    Building an Advanced Drawing Machine, 95–98
    Building with Bioglue, 67–68
    Candle-Powered Boat, 77–79
    Cell Phone Signal Blocking, 59–60
    Chaos and the Double Pendulum, 15–16
    Chlorophyll Printing, 99–101
    Creating Time, 113–115
    Design a Hologram, 105–106
    Dissecting Sound, 35–37
    Floating Dollar Bills, 24–25
    Geological Art, 102–103
    Homopolar Motor Color Whirl, 61–63
    Hydroponic System, 51–52
    Icosahedral Virus, 118–119
    Insect Mind Control, 7–8
    Light Versus Sound, 48–49
    Make a Knockoff Ultraviolet Light, 53–55
    Make a Microscope, 43–44
    Momentum Drop Painting, 93–94
    The Original Fidget Spinner, 38–39
    Put a Cork in It, 75–76
    Rock-Hopping Robot, 83–85
    Rotational Races, 29–31
    Rubber Band Power 2-in-1, 33–34
    Secret Patterns of Sound Waves, 45–47
    Seeing the World Through a Bubble, 12–13
    Shooting Straws, 125–127
    Shrinking Plastic, 27–28
    Spaghetti Strength, 81–82
    Static UFO, 9–11
    Super Simple Stethoscopes, 56–57
    Using the Golden Ratio to Make a Mechanical
        Tentacle, 111–112
    Walk-Along Spinning Fish, 72–73

Technology

    Building 3D Lenses, 21–23
    Building an Advanced Drawing Machine, 95–98
    Candle-Powered Boat, 77–79
    Cell Phone Signal Blocking, 59–60
    Chlorophyll Printing, 99–101
    Complex Counting Challenge, 120–121
    Creating Time, 113–115
    Design a Hologram, 105–106
    Dissecting Sound, 35–37
    Homopolar Motor Color Whirl, 61–63
    Hydroponic System, 51–52
    Light Versus Sound, 48–49
    Make a Knockoff Ultraviolet Light, 53–55
    Make a Microscope, 43–44
    Make Your Own Molds, 69–71
    Put a Cork in It, 75–76
    Rock-Hopping Robot, 83–85
    Secret Patterns of Sound Waves, 45–47
    Series Versus Parallel String Light Circuits, 87–88
    Shrinking Plastic, 27–28
    Super Simple Stethoscopes, 56–57
    Walk-Along Spinning Fish, 72–73
    Weaving a Magic Square, 116–117

INDEX

A

Acoustical engineering, 45–47
Additive light theory, 61–63
Aeronautics, 124
Air pressure, 72–73
Anaglyphs, 21–23
Anemometers, 16
Angular momentum, 31
Anthotypes, 99–101
Art, 91

B

Barycenter, 25
Bases, of number systems, 120–121
Beam bridges, 81–82
Bell, Alexander Graham, 124
Benchmarks, 109
Bestie Marble Maze Test, 17–19
Binary number system, 120–121
Bioglues, 67–68
Biomimicry, 68
Brand, Matthew, 106
Bridges, 81–82
Building 3D Lenses, 21–23
Building an Advanced Drawing Machine, 95–98
Building with Bioglue, 67–68
Butterfly effect, 15–16

C

Candle-Powered Boat, 77–79
Casts, 71
Cell Phone Signal Blocking, 59–60

Center of gravity, 24–25
Center of mass, 25
Centripetal force, 38–39
Chaos and the Double Pendulum, 15–16
Chaos theory, 15–16
Chemical reactions, 75–76
Chemical weathering, 103
Chemoreceptors, 8
Chladni, Ernst, 47
Chladni Plates, 47
Chlorophyll Printing, 99–101
Circuits, 87–88
Collaboration, 129
Colors, 12–13, 61–63
Communication, 129
Complex Counting Challenge, 120–121
Convection currents, 77–79
Creating Time, 113–115
Creativity, 129
Critical thinking, 129

D

Decimal number system, 120
Deposition, 103
Depth perception, 105–106
Design a Hologram, 105–106
Design thinking, viii, 57, 91
Destructive interference, 12–13
Dissecting Sound, 35–37
Dürer, Albrecht, 117
Dynamics, 15–16

E

Electric and magnetic fields (EMFs), 60
Electromagnetic pulses (EMPs), 60
Electromagnetic radiation, 59–60
Electrostatic force, 9–11
Energy, 33–34
Engineering, 65
Engineering design process, 65
Entomologists, 7–8
Erosion, 103
Experimenting, 3

F

Failure, 3
Faraday cages, 59–60
Floating Dollar Bills, 24–25
Fluorescence, 53–55
Force, centripetal, 38–39
"Four Cs," 129
Frequency, 35–37, 45–47
Friction, 95–98

G

Geological Art, 102–103
Ginormous Tetrahedral Challenge, 123–124
Golden ratio, 111–112
Gravity, center of, 24–25

H

Hand-eye coordination, 17–19
Harmonic motion, 95–98
Harmonographs, 95–98
Heart rate, 56–57
Hertz (Hz), 47
Holograms, 105–106
Homopolar Motor Color Whirl, 61–63
Hydroponic System, 51–52
Hypotheses, 2–3

I

Icosahedral Virus, 118–119
Incident angles, 48–49
Inertia, 29–31, 38–39

Insect Mind Control, 7–8
Interdisciplinary, 1

K

Kinetic energy, 33–34

L

Laënnec, René, 57
Law of Reflection, 48–49
Light Versus Sound, 48–49
Light waves, 12–13, 48–49

M

Magic constants, 116–117
Magic squares, 116–117
Make a Knockoff Ultraviolet Light, 53–55
Make a Microscope, 43–44
Make Your Own Molds, 69–71
Math, 109
Microorganisms, 43–44
Microscopes, 43–44
Models, 71
Molds, 71
Moment of Inertia, 29–31
Momentum, 31, 93–94
Momentum Drop Painting, 93–94
Motors, homopolar, 61–63

N

National Academy of Engineering (NAE), 65
Newton's Third Law, 75–76
Notebooks, 2
Number systems, 120–121

O

The Original Fidget Spinner, 38–39

P

Parallax, 23
Parallel circuits, 87–88
Pendulums, 15–16, 95–98, 113–115
Periods (time), 113–115
Perspective, 21–23, 105–106

Pheromones, 7–8
Photosynthesis, 101
Physical weathering, 103
Pigments, 99–101
Pitch, 35–37
Pollock, Jackson, 94
Polyhedrons, 118–119
Polystyrene, 27–28
Potential energy, 33–34
Printing, 99–101
Projectiles, 125–127
Protists, 43–44
Put a Cork in It, 75–76

R
Radio waves, 59–60
Reflected angles, 48–49
Rock-Hopping Robot, 83–85
Rotational Races, 29–31
Rubber Band Power 2-in-1, 33–34

S
Safety rules, 2
Science, 5
Scientific method, ix
Scratch holography, 105–106
Secret Patterns of Sound Waves, 45–47
Seeing the World Through a Bubble, 12–13
Self-folding plastics, 28
Series circuits, 87–88
Series Versus Parallel String Light Circuits, 87–88
Shooting Straws, 125–127
Shrinking Plastic, 27–28
Sound waves, 35–37, 45–47, 48–49

Spaghetti Strength, 81–82
Spatial navigation, 17–19
Static UFO, 9–11
STEAM acronym, viii–ix, 1. *See also* Art; Engineering; Math; Science; Technology
Stethoscopes, 56–57
Super Simple Stethoscopes, 56–57
Synergy, 129

T
Technology, 41
Tetrahedrons, 123–124
Thermoplastics, 27–28
Three-dimensional (3D) images, 21–23
Torque, 83–85

U
Ultraviolet (UV) light, 53–55
Using the Golden Ratio to Make a Mechanical Tentacle, 111–112

V
Variables, 5
Velocity, 93–94
Viruses, 118–119

W
Walk-Along Spinning Fish, 72–73
Waveform visualizers, 48–49
Weaving a Magic Square, 116–117
Wheels, 83–85
Wick systems, 51–52
Wind, 77–79

ACKNOWLEDGMENTS

This book would not be possible without the help and inspiration of many people.

First and foremost, a huge thank you to my best friend and husband, Adam. Thank you for your excitement in doing the experiments in this book with me. Thank you for reading through my rough drafts with your engineer's sensibility. Thank you for taking our daughter on adventures while I stayed home writing this book. Most of all, thank you for your never-ending support for all that I do.

Thank you to all my family. Thank you, Mom, for all the nights you let me stay up late reading and the endless concoctions you allowed us kids to make in the kitchen. Thank you, G-ma, for all the library visits and lunch dates while I was in school. Thank you, Dad, for being a passionate steward of our environment and picking up all the trash everywhere we go. It used to embarrass me, but now I see how important it is for everyone to protect our planet. The next book will be *our* book. Thank you to my siblings, Matt, Corrie, Derrek, and Sarah, who continue to inspire me to

be better. And of course, thank you to my daughter, Ada, who helps me learn something new every day. I love you all.

Thank you to all the teachers in my life, including my fifth grade teacher, Mrs. Sauceda, who inspired me to pursue my own path. A special thanks to ALL the teachers and teacher leaders that I have had the privilege of working with, especially Susanne, and of course, all my B-hall family: Michelle, Bobby, Chrystle, Kevin, Carol, Melanie, Amber, Dove, Lori, Marty, and Sandie. Thank you to all my students. You motivate me to be the best teacher I can be.

Thank you to my editors, Jeanine Le Ny and Deb Housel, and all the staff at Callisto for helping make this book the best it can be. Special thanks to Joe Cho and Susan Randol for making my dream of writing a book a reality.

Finally, to you, the readers of this book, thank you for sharing a love for STEAM. I hope this book challenged you and motivated you to learn new things.

ABOUT THE AUTHOR

Jess Harris has ten years of experience as a public school teacher. She spent the first five years as an elementary educator teaching all subjects to fifth graders. She spent the next five years as a high school science educator teaching earth and environmental science, physics, physical science, and AP Biology. She has a master's degree in science education from East Carolina University and is a Nationally Board Certified Teacher for Adolescence and Young Adulthood in Science. She lives in North Carolina with her husband, daughter, and two cats. She loves sharing about science on her website, MrsHarrisTeaches.com.